T0237753

Lecture Notes in Bioinformatics

Edited by S. Istrail, P. Pevzner, and M. Waterman

Subseries of Lecture Notes in Computer Science

Aoife McLysaght   Daniel H. Huson (Eds.)

Comparative Genomics

RECOMB 2005 International Workshop, RCG 2005
Dublin, Ireland, September 18-20, 2005
Proceedings

Series Editors

Sorin Istrail, Celera Genomics, Applied Biosystems, Rockville, MD, USA
Pavel Pevzner, University of California, San Diego, CA, USA
Michael Waterman, University of Southern California, Los Angeles, CA, USA

Volume Editors

Aoife McLysaght
University of Dublin, Smurfit Institute of Genetics, Trinity College, Irland
E-mail: mclysaga@tcd.ie

Daniel H. Huson
Tübingen University, Center for Bioinformatics (ZBIT)
72076 Tübingen, Germany
E-mail: huson@informatik.uni-tuebingen.de

Library of Congress Control Number: 2005932241

CR Subject Classification (1998): F.2, G.3, E.1, H.2.8, J.3

ISSN   0302-9743
ISBN-10   3-540-28932-1 Springer Berlin Heidelberg New York
ISBN-13   978-3-540-28932-6 Springer Berlin Heidelberg New York

Springer is a part of Springer Science+Business Media

springeronline.com

© Springer-Verlag Berlin Heidelberg 2005
Printed in Germany

Typesetting: Camera-ready by author, data conversion by Scientific Publishing Services, Chennai, India
Printed on acid-free paper   SPIN: 11554714   06/3142   5 4 3 2 1 0

Preface

The complexity of genome evolution poses many exciting challenges to developers of mathematical models and algorithms, who have recourse to a spectrum of algorithmic, statistical and mathematical techniques, ranging from exact, heuristic, fixed-parameter and approximation algorithms for problems based on parsimony models to Monte Carlo Markov Chain algorithms for Bayesian analysis of problems based on probabilistic models.

The annual RECOMB Satellite Workshop on Comparative Genomics (RECOMB Comparative Genomics) is a forum on all aspects and components of this field, ranging from new quantitative discoveries about genome structure and process to theorems on the complexity of computational problems inspired by genome comparison. The informal steering committee for this meeting consists of David Sankoff, Jens Lagergren and Aoife McLysaght.

This volume contains the papers presented at the 3rd RECOMB Comparative Genomics meeting, which was held in Dublin, Ireland, on September 18–20, 2005. The first two meetings of this series were held in Minneapolis, USA (2003) and Bertinoro, Italy (2004).

This year, 21 papers were submitted, of which the Program Committee selected 14 for presentation at the meeting and inclusion in this proceedings. Each submission was refereed by at least three members of the Program Committee. After completion of the referees' reports, an extensive Web-based discussion took place for making decisions. The RECOMB Comparative Genomics 2005 Program Committee consisted of the following 27 members: Vineet Bafna, Anne Bergeron, Mathieu Blanchette, Avril Coghlan, Dannie Durand, Nadia El-Mabrouk, Niklas Eriksen, Aaron Halpern, Rose Hoberman, Daniel Huson, Jens Lagergren, Giuseppe Lancia, Emmanuelle Lerat, Aoife McLysaght, Istvan Miklos, Bernard Moret, Pavel Pevzner, Ben Raphael, Marie-France Sagot, David Sankoff, Cathal Seoighe, Beth Shapiro, Igor Sharakhov, Mike Steel, Jens Stoye, Glenn Tesler and Louxin Zhan. We would like to thank the Program Committee members for their dedication and hard work.

RECOMB Comparative Genomics 2005 had several invited speakers, including: Anne Bergeron (Université du Québec à Montreal, Canada), Laurent Duret (Laboratoire de Biometrie et Biologie Evolutive, Université Claude Bernard, Lyon, France), Eddie Holmes (Department of Biology, Pennsylvania State University, USA), Jeffrey Lawrence (Department of Biological Sciences, University of Pittsburgh, USA), Stephan Schuster (Department of Biochemistry and Molecular Biology, Pennsylvania State University, USA), Ken Wolfe (Genetics Department, Trinity College Dublin, Ireland) and Sophia Yancopoulos (Institute for Medical Research, New York, USA).

In addition to the invited talks and the contributed talks, an important ingredient of the program was the lively poster session.

RECOMB Comparative Genomics 2005 would like to thank Science Foundation Ireland (SFI) and Hewlett-Packard for providing financial support for the conference. We would like to thank the University of Dublin, Trinity College, for hosting the meeting. We would like to thank Nadia Browne for administrative support.

In closing, we would like to thank all the people who submitted papers and posters and those who attended RECOMB Comparative Genomics 2005 with enthusiasm.

September 2005   Aoife McLysaght and Daniel Huson

Table of Contents

Lower Bounds for Maximum Parsimony with Gene Order Data
  *Abraham Bachrach, Kevin Chen, Chris Harrelson, Radu Mihaescu,
  Satish Rao, Apurva Shah* 1

Genes Order and Phylogenetic Reconstruction: Application to
γ-Proteobacteria
  Guillaume Blin, Cedric Chauve, Guillaume Fertin 11

Maximizing Synteny Blocks to Identify Ancestral Homologs
  Guillaume Bourque, Yasmine Yacef, Nadia El-Mabrouk 21

An Expectation-Maximization Algorithm for Analysis of Evolution of
Exon-Intron Structure of Eukaryotic Genes
  Liran Carmel, Igor B. Rogozin, Yuri I. Wolf, Eugene V. Koonin 35

Likely Scenarios of Intron Evolution
  Miklós Csűrös 47

OMA, a Comprehensive, Automated Project for the Identification
of Orthologs from Complete Genome Data: Introduction and First
Achievements
  *Christophe Dessimoz, Gina Cannarozzi, Manuel Gil,
  Daniel Margadant, Alexander Roth, Adrian Schneider,
  Gaston H. Gonnet* 61

The Incompatible Desiderata of Gene Cluster Properties
  Rose Hoberman, Dannie Durand 73

The String Barcoding Problem is NP-Hard
  Marcello Dalpasso, Giuseppe Lancia, Romeo Rizzi 88

A Partial Solution to the C-Value Paradox
  Jeffrey M. Marcus 97

Individual Gene Cluster Statistics in Noisy Maps
  Narayanan Raghupathy, Dannie Durand 106

Power Boosts for Cluster Tests
  David Sankoff, Lani Haque 121

Reversals of Fortune
  David Sankoff, Chungfang Zheng, Aleksander Lenert 131

Very Low Power to Detect Asymmetric Divergence of Duplicated Genes
Cathal Seoighe, Konrad Scheffler 142

A Framework for Orthology Assignment from Gene Rearrangement Data
Krister M. Swenson, Nicholas D. Pattengale, Bernard M.E. Moret ... 153

Author Index 167

Lower Bounds for Maximum Parsimony with Gene Order Data

Abraham Bachrach, Kevin Chen, Chris Harrelson, Radu Mihaescu,
Satish Rao, and Apurva Shah

Department of Computer Science,
UC Berkeley

Abstract. In this paper, we study lower bound techniques for branch-and-bound algorithms for maximum parsimony, with a focus on gene order data. We give a simple $O(n^3)$ time dynamic programming algorithm for computing the maximum circular ordering lower bound, where n is the number of leaves. The well-known gene order phylogeny program, GRAPPA, currently implements two heuristic approximations to this lower bounds. Our experiments show a significant improvement over both these methods in practice. Next, we show that the linear programming-based lower bound of Tang and Moret (Tang and Moret, 2005) can be greatly simplified, allowing us to solve the LP in $O^*n^3)$ time in the worst case, and in $O^*(n^{2.5})$ time amortized over all binary trees. Finally, we formalize the problem of computing the circular ordering lower bound, when the tree topologies are generated bottom-up, as a *Path-Constrained Traveling Salesman Problem*, and give a polynomial-time 3-approximation algorithm for it. This is a special case of the more general *Precedence-Constrained Travelling Salesman Problem* and has not previously been studied, to the best of our knowledge.

1 Introduction

Currently, the most accurate methods for phylogenetic reconstruction from gene order data are based on branch-and-bound search for the most parsimonious tree under various distance measures. These include GRAPPA [1], BP-Analysis [2], and the closely-related MGR [3]. Since branch-and-bound for this problem is potentially a super-exponential-time process, computing good pruning lower bounds is very important. However, scoring a particular partial or full tree topology is a hard computational problem for many metrics, in particular for gene order data.

There has been a good deal of recent work on designing good lower bounds for various distance measures. These techniques are divided between those specially designed for specific distance measures [4–6] and those that hold for arbitrary metrics [7–9]. Our lower bounds fall into the latter category. Lower bounds that hold for arbitrary metrics are particularly appealing in the context of gene order phylogeny, because an important direction for the field is to extend current methods to use more realistic metrics than the breakpoint or inversion distances currently used. There is a growing body of algorithmic work on various

A. McLysaght et al. (Eds.): RECOMB 2005 Ws on Comparative Genomics, LNBI 3678, pp. 1–10, 2005.

distance measures, including transpositions, chromosome fusions/fissions, insertions/deletions and various combinations of these (see [10] for a comprehensive survey) and our lower bounds apply to all of them. One notable exception to this is the tandem duplication and random loss model [11], which is well-suited to animal mitochondrial genomes, but is asymmetric and therefore does not fit into the standard metric parsimony framework.

In this paper, we give efficient implementations of two lower bounds. The first is a simple dynamic programming algorithm to compute the maximum circular ordering lower bound in $O(n^3)$ time and $O(n^2)$ space. Since the exact running time of the algorithm often depends on the choice of root, we also provide an algorithm to compute the optimal root for a given un-rooted tree topology in $O(n^2)$ time. Next, we greatly simplify the LP-based lower bound of [9] and show how to implement it in $O^*(n^3)$ time[1] in the worst case and $O^*(n^{2.5})$ time amortized over all binary trees. Finally, we study the problem of lower bounding the tree score when the only a partial topology has been constructed so far and rephrase this as a *Path-Constrained Travelling Salesman Problem*. This is a special case of the *Precedence-Constrained Travelling Salesman Problem* [12], in which we are given a partial order graph on a subset of the cities and asked to return a min cost tour that respects the partial ordering. Our version of the problem is simply the case where the partial order graph is a directed path. To our knowledge, considering the effect of a restricted partial order on this problem has not been previously studied, and we give a simple and fast algorithm that computes a 3-approximation for the case of a line. The solution can then be transformed into a lower bound by dividing the score by 3.

Finally, we have implemented our dynamic programming lower bound and show that it gives better results on the benchmark Campanulaceae data set.

2   The Circular Ordering Lower Bound

Given a rooted binary tree in which one of each pair of children is designated the left child and the other the right child, consider the left-to-right ordering of the leaves, π, induced by some depth-first search of the tree. For a given metric $d(\cdot)$ on pairs of leaves, say the inversion distance, we define the circular ordering lower bound to be $C(\pi) = \sum_{i=1}^{n} d(\pi(i), \pi(i+1))$, where we define $\pi(n+1) = \pi(1)$ for notational convenience. By repeatedly invoking the triangle inequality, it is easy to see that $\frac{C(\pi)}{2}$ is a lower bound on the cost of the tree, and the bound is tight if the distance $d(\cdot)$ is the shortest path metric induced by the tree.

The same tree topology can induce more than one leaf ordering π by swapping left child and right children at internal nodes of the tree, and some leaf orderings may produce a higher lower bound that others. A brute-force exponential time algorithm for computing the maximum circular ordering is to enumerate all possible combinations of swaps at internal nodes. This method has been considered too expensive to work well in practice [13]. Two heuristic approximations

---
[1] The notation $O^*(f(n))$ omits factors that are logarithmic in n.

are implemented in GRAPPA. The first is the *Swap-as-you-go* heuristic [7], in which a DFS traversal of the tree is performed, and a swap performed at each internal node when it is visited as long as it improves the lower bound. This heuristic has the attribute of running in linear time.

The second heuristic does a similar traversal, but when deciding whether to swap the left and right children of an internal node, the algorithm tries both of them, and keeps the one which gives the better lower bound for the current subproblem. This latter approach runs in $O(n^4)$ time.

In this section, we show that, in fact, the maximum cicular ordering for a given tree can be computed in $O(n^3)$ time by a straightforward dynamic programming algorithm. We first note that the choice of the tree's root does not affect the parsimony score or the maximum circular ordering lower bound. Since the branch-and-bound algorithm searches over unrooted topologies, for the presentation of the algorithm we assume an arbitrarily chosen root. On the other hand, the exact running time of our algorithm will depend on the position of the root, so we also consider the problem of optimal root placement after giving the description of the algorithm.

At each internal node, v, let S_v be the set of leaves in the subtree rooted at v. The dynamic programming algorithm constructs a table M_v which contains, for each pair of leaves $A, B \in S_v$, the maximum score attainable by a *linear* ordering of the vertices in S_v that begins with A and ends with B, if one exists. Note that such an ordering exists if and only if A and B are leaves in subtrees rooted at different children of v.

Let the children of v be l and r. We inductively assume that the tables $M(r)$ and $M(l)$ have already been constructed. Let ll and lr be l's children and rl and rr be r's children, and let us assume that the subtrees rooted at ll, lr, rl and rr have a, b, c and d leaves respectively (if r or l are leaves, then this step may be omitted). Intuitively, we could construct $M(v)$ by considering all possible quartets of leaves $A \in S_{ll}, B \in S_{lr}, C \in S_{rl}$ and $D \in S_{rr}$. We will then perform $O(abcd)$ operations at node v, which would lead to a running time of $O(n^4)$ for the whole tree.

We can do better than this naive implementation in the following way. Let $A \in S_{ll}$ and $C \in S_{rl}$. Let

$$\delta(A, C) = \max_{B \in S_{lr}} [M_l(A, B) + d(B, C)]. \tag{1}$$

So $\delta(A, C)$ is the highest score attainable by a linear ordering of the leaves in S_l which also takes a final step to C. Now for $D \in S_{rr}$ we obtain

$$M_v(A, D) = \max_{C \in S_{rl}} [\delta(A, C) + M_r(C, D)]. \tag{2}$$

The maximum circular ordering lower bound is given at the root by the expression

$$\max_{A \in S_l, D \in S_r} [M_v(A, D) + d(A, D)]$$

2.1 Analyzing the Running Time

To analyze the running time of the algorithm, assume inductively that the time to process an n-leaf tree is $O(n^3)$. Keeping the notation as before, at a given node v, by induction, it takes $O((a+b)^3)$ time to compute the table M_l and $O((c+d)^3)$ time to compute M_r. We need to show that the time to compute the table M_v is $O((a+b+c+d)^3)$, since $a+b+c+d$ is the number of leaves in the subtree rooted at v.

To complete the computation of equation 1 for every pair $A \in S_{ll}, C \in S_{rl}$ (there are ac of them), we need $O(b)$ time and to complete the computation in Equation 2 for all ad pairs $A \in S_{ll}, D \in S_{rr}$ we need $O(c)$ time per pair. The running time to compute the entries $M_v(A, D)$ with $A \in S_{ll}$ and $D \in S_{rr}$ is therefore $O(abc + acd)$. Of course we need to do this for all possible choices of subtrees of l and r, therefore the total running time at node v will be $O(abc + acd + abd + bcd) = O((a+b+c+d)^3)$.

Observe that at each step we are keeping one table of size $O(ab)$ for a node with a leaves under its left child and b leaves under its right child. A similar identical to the one above proves that the space needed for building a tree on n leaves is at most $O(n^2)$. This proves the following theorem:

Theorem 1. *There exists an $O(n^3)$-time, $O(n^2)$-space algorithm for computing the maximum circular ordering lower bound.*

The worst case time and space are achieved by a balanced binary tree. Conversely, the best running time for the algorithm is achieved when the tree is a caterpillar. In this case, the running time is only $O(n^2)$.

Up to now, we have placed the root arbitrarily. However, observe that the choice of the root can affect the running time of the algorithm. For example, consider a rooted tree consisting of a root node with four grandchildren, where

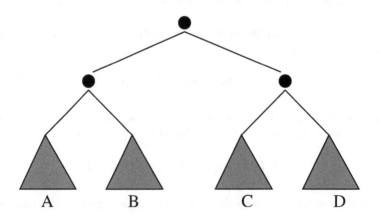

Fig. 1. Each of the shaded triangles A, B, C represent a caterpillar on $O(n)$ leaves, while D is the singleton tree

three of the four subtrees rooted at the grandchildren are caterpillars with n leaves (Figure 1) and the fourth one is a singleton leaf.

The running time for this rooting is $\Omega(n^3)$. However, if we root the tree somewhere inside C, then the running time becomes $O(n^2)$.

The optimal choice of root can be found easily in $O(n^2)$ time simply by considering each of the possible root placements in turn, and traversing the tree in depth-first search order, starting at each root, to compute the running time that the algorithm would need if that particular rooting were chosen. Since this algorithm's running time s $O(n^2)$, we can run it first to choose the root without increasing the overall asymptotic running time of $O(n^3)$.

It is easily seen that for a balanced binary tree, one can never do better than $\Omega(n^3)$ regardless of the rooting. However, it is not clear what is the average running time of the algorithm for an optimal rooting across the set of all trees. We leave answering this question as an open problem.

3   An LP-Based Lower Bound

A recent paper by Tang and Moret gives an alternative lower bound for general metrics based on linear programming [9]. In their LP, there is a variable x_e for each tree edge, e, which corresponds to the length of the edge, and a variable $y_{i,j}$ for each pair of non-leaf vertices i, j at distance 2 in the tree (in terms of edges), corresponding to the distance between these two vertices. If i and j are leaves, we define $y_{i,j} = d(i,j)$.

The object is to minimize the sum of the x_e subject to triangle inequality constraints and *perfect median* constraints. There is one perfect median constraint for each internal vertex v. Let i, j, k be the three neighbors of this vertex. The sum of the edge lengths of the three edges adjacent to v is constrained to be *equal* to $\frac{y_{i,j}+y_{j,k}+y_{k,i}}{2}$, which essentially says that the circular ordering lower bound for the local subtree around each internal vertex is tight - hence the term *perfect median*. The motivation for this constraint is an empirical observation that in many practical situations, many of the perfect median constraints are indeed satisfied. In addition to these constraints, there are also two triangle inequality constraints for each pair of leaves i, j, in which two paths made up of sums of x and y variables between the leaves are constrained to be at least the measured distance between the two leaves, $d(i,j)$ (the choice of the two paths is not deterministically given in the paper).

In this section, we show that this LP is equivalent to the following very simple LP, which has appeared previously in the literature (e.g. [14]). We do away completely with the y variables and keep only the x variables. Let L be the set of leaves of the tree, $path(i,j)$ be the (unique) shortest path from i to j in the tree, and $d(i,j)$ be the measured distance between i and j (e.g., $d(i,j)$ could be the inversion distance between i and j). The LP, which we call the *triangle inequality LP* is:

$$min \sum_{e \in E} x_e$$

$$subject\ to$$

$$\forall i,j \in L \sum_{e \in path(i,j)} x_e \geq d_{ij}$$

$$\forall e\ x_e \geq 0$$

Theorem 2. *The LP of Tang and Moret is equivalent to the triangle inequality LP.*

Proof. To show equivalence between the two LP's, we need to prove that all constraints in the LP of Tang and Moret that involve y variables can *always* be satisfied by a solution that satisfies all the constraints that involve only x variables (i.e. the triangle inequality LP). We consider two cases separately: first, at interior nodes with at most one neighbor that is a leaf, second, at interior nodes with exactly two neighbors that are leaves. For trees with more than three leaves, there are no other cases to consider, and for a tree with three leaves the theorem is trivial.

- Case 1: Given a solution to the triangle inequality LP, at an interior node v with neighbors i, j, k, none of which are leaves, we can just set the y_{ij} variable to be the sum of $x_{\{i,v\}}$ and $x_{\{j,v\}}$, and likewise for y_{jk} and y_{ki}. This satisfies both the perfect median constraint for this interior node and any triangle inequality constraints involving y_{ij}, y_{jk} or y_{ki}.
  If one of the tree is a leaf, the same argument holds, since the triangle inequality with respect to two leaves cannot be violated.
- Case 2: At an interior node v with adjacent leaves i, j and a third adjacent node k, this may not be possible to proceed as in case 1 in general, because $d(i,j)$ could be smaller than the sum of $x_{\{i,v\}}$ and $x_{\{j,v\}}$. But to compensate, we can simply increase the other two y variables by the appropriate amount in order to satisfy the perfect median constraint, and this will still satisfy the relevant triangle inequality constraints. Note that this would in principle violate local triangle inequality constraints between i and k and between j and k, but these constraints are not included in the LP of Tang and Moret. (If they were included, then the LP would not be satisfiable by the most parsimonious tree, so its use as a lower bound would be in question.)   □

Next we observe that our simpler LP has another attribute: it has the form of a pure covering LP, for which many efficient approximation schemes exist. For example, we can apply the algorithm of [15], which solves the LP to within a factor of $1 + \epsilon$ in time $O^*(\frac{1}{\epsilon^2}m)$ where m is the number of non-zero entries in the constraint matrix. We can always divide the solution by $1 + \epsilon$ to get a lower bound on the tree score. We further note that the algorithm of [15] is quite simple and practical to implement, and does not require any fancy linear programming machinery.

Note that m is the sum of the path lengths between all pairs of leaves in the tree and ranges from $n^2 \log(n)$ for a complete binary tree to n^3 for a caterpiller. In general, m can be bounded by the product of the number of constraints, n^2, and the height of the tree. Using a classical result of Flajolet and Odlyzko, which states that the height of a random binary tree is $O(\sqrt{n})$ [16], also observe that solving the LP takes $O^*(n^{2.5})$ per tree, amortized over all trees with n leaves. These results are summarized in the following theorem.

Theorem 3. *The linear program 3 can be solved in $O^*(n^3)$ time in the worst case and $O^*(n^{2.5})$ time amortized over all trees.*

Also, note that while the LP-based algorithm is fastest on balanced trees, the dynamic programming algorithm is fastest on unbalanced trees. Hence, we can optimize the running time (although not asymptotically) by first computing the running time for the two algorithms and running the lower of the two.

4   The Path-Constrained Traveling Salesman Problem

The current implementation of GRAPPA generates each possible full tree topology and computes the circular ordering lower bound for it in order to determine whether to proceed with a full scoring. This strategy can be improved by generating trees bottom-up instead, inserting leaves one at a time into a partial topology, and computing the circular ordering lower bound on the partial topology. It is easy to see that a lower bound on a partial topology is a valid lower bound on any extension of it into a full tree. In this way, entire subtrees of the branch-and-bound recursion tree can be pruned at an earlier stage.

When such a bottom-up strategy is adopted, the problem of computing a circular ordering that is consistent with the partial tree can be rephrased as a *Precedence-Constrained Travelling Salesman Problem*. Here, we are asked to produce a min-cost tour of a set of cities that respect a set of precedence constraints in the form of a directed acyclic graph (DAG). This problem has been studied in [12], in which hardness results were given for the special cases of metrics induced by a line or hypercube, suggesting that the problem is hard in many practical instances.

For our application, we are interested in specializing not the metric, but the constraint graph. In contrast to [12], our special case is easy to approximate. In general, the problem of computing a maximum circular ordering lower bound can be seen as a version of TSP. Given a partial topology, a maximum linear ordering on this topology induces a constraint graph in which the DAG is a directed path. Clearly, Path-Constrained TSP is NP-hard since it contains TSP as a special case. However, we are able to prove the following result:

Theorem 4. *There exists an 3-approximation for the path-constrained travelling salesman problem that runs in the same time as computing a minimum spanning tree on the given metric on the leaves.*

Proof. We will call the edges and vertices of the constraint graph *constraint edges* and *constraint vertices* respectively. First, take the complete graph on the

n vertices with the distance $d(i, j)$ on the edge from i to j and form a new graph by adding an auxiliary vertex x with 0-weight edges connected to each of the constraint vertices. Find a minimum spanning tree in this new graph. The optimal solution will contain the 0-weight edges, plus a tree growing from each of the constraint vertices. Discarding the 0-weight edges and the auxiliary vertex x, we have a forest where each tree contains exactly one of the constraint vertices. We produce a travelling salesman tour of each tree by following an Euler tour of the spanning tree. The sum of the tree costs is a lower bound on OPT because the optimal solution must have a path between consecutive constraint vertices, and removing the last edge on each of these paths produces a forest in which each tree contains exactly one of the constraint vertices. Therefore the sum of the costs of these tours is at most twice OPT. The final traveling salesman tour of the whole graph combines the individual tours with the (undirected) constraint edges, with short-cutting if necessary. By the triangle inequality, the sum of all of the path edges is a lower bound on OPT, so the final solution is a 3-approximation to the optimal path-constrained TSP.   □

This approximation ratio implies a lower bound for the branch-and-bound algorithm by taking the cost of the solution and dividing by 3. We also note that the well-known Christofides technique of adding a min cost matching on vertices of odd degree does not seem to be easily applicable in our problem.

5   Experiments

We ran our maximum circular-ordering dynamic programming algorithm on a benchmark set of 12 chloroplast genomes from the Campanulaceae family (see Table 5) on a workstation with 1Gb of memory. The lower bound strategy implemented in GRAPPA is to try the default circular ordering first, then the two above-mentioned heuristics, stopping at the first time a lower bound that exceeds the current best score is found. Our implementation adds our dynamic programming algorithm for computing the max circular ordering lower bound as an additional bound if the first two bounds are not good enough. Our results show that we get a significant improvement in both number of trees scored and in total running time.

These experiments used the naive $O(n^4)$ implementation of the dynamic programming so the running time would be even faster for the $O(n^3)$ implementation. In practice, it may be possible to improve the running time slightly by

	Camp 10		Camp 12	
	Trees scored	Time (min)	Trees scored	Time (min)
GRAPPA	6,391	3:40	80,551	165
Our algorithm	4,277	2:37	42,124	102

Fig. 2. Camp 12 is the full data set of 12 chloroplast genomes and Camp 10 is a subset of 10 of those genomes. Both data sets are distributed with the GRAPPA source code.

sharing partial computations between different trees, since many subtrees are shared between trees.

6   Discussion

It has recently come to our attention that the $O(n^3)$ result for the maximum circular ordering problem (Section 2) was previously achieved in two completely different contexts [17,18].

We also remark that the problem we consider in this paper is related to the classic *Distance Wagner Method*, in which the problem is to compute the shortest tree that dominates a given distance matrix [19]. Our problem differs in that we are also given the tree topology as an input.

Acknowledgements

We thank Bernard Moret for helpful discussions and Jijun Tang for sharing his code with us. We also thank the three anonymous reviewers for pointing us to references [17,18,19,14]. This work was supported by NSF grant EF 03-31494. Radu Mihaescu was also supported by the Fannie and John Hertz Foundation.

References

1. B. Moret, S. Wyman, D. Bader, T. Warnow, and M. Yan. A new implementation and detailed study of breakpoint analysis. In *PSMB*, 2001.
2. D. Sankoff and M. Blanchette. Multiple genome rearrangement and breakpoint phylogeny. *J. Comput. Biol.*, 5(3):555–570, 1998.
3. G. Bourque and P. Pevzner. Genome-scale evolution: Reconstructing gene orders in the ancestral species. *Genome Res.*, 12(1):26–36, 2002.
4. D. Bryant. A lower bound for the breakpoint phylogeny problem. *Journal of Discrete Algorithms*, 2:229–255, 2004.
5. P. Purdom, P. Bradford, K. Tamura, and S. Kumar. Single column discrepency and dynamic max-mini optimizations for quickly finding the most parsimonious evolutionary trees. *Bioinformatics*, 16(2):140–151, 2000.
6. B. Holland, K. Huber, D. Penny, and V. Moulton. The minmax squeeze: Guarenteeing a minimal tree for population data. *Mol. Biol. and Evol.*, 22(2):235–242, 2005.
7. J. Tang. A study of bounding methods for reconstructing phylogenies from gene-order data. PhD Thesis, 2003.
8. J. Tang, B. Moret, L. Cui, and C. dePamphilis. Phylogenetic reconstruction from arbitrary gene-order data. In *BIBE*, 2004.
9. J. Tang and B. Moret. Linear programming for phylogenetic reconstruction based on gene rearrangements. In *CPM*, 2005.
10. B. Moret, J. Tang, and T. Warnow. Reconstructing phylogenies from gene-content and gene-order data. In O Gascuel, editor, *Mathematics of Evolution and Phylogeny*. Oxford Univ. Press, 2004.
11. K. Chaudhuri, K. Chen, R. Mihaescu, and S. Rao. On the tandem duplication-random loss model of genome rearrangement. In review.

12. M. Charikar, R. Motwani, P. Raghavan, and C. Silverstein. Constrained TSP and low power computing. In *WADS*, 1997.
13. B. Moret. Personal communication, 2005.
14. G. Lancia and R. Ravi. GESTALT: Genomic steiner alignments. In *CPM*, 1999.
15. N. Young. Sequential and parallel algorithms for mixed packing and covering. In *FOCS*, 2001.
16. P. Flajolet and A. M. Odlyzko. The average height of binary trees and other simple trees. *J. Computer System Sci.*, 25:171–213, 1982.
17. Z. Bar-Joseph, E. D. Demaine, D. K. Gifford, A. M. Hamel, T. S. Jaakkola, et al. K-ary clustering with optimal leaf ordering for gene expression data. *Bioinformatics*, 19(9):1070–1078, 2003.
18. R. E. Burkard, V. G. Deineko, and G. J. Woeginger. The travelling salesman and the pq-tree. *Mathematics of Operations Research*, 24:262–272, 1999.
19. M. Farach, S. Kannan, and T. Warnow. A robust model for finding optimal evolutionary trees. In *STOC*, 1993.

Genes Order and Phylogenetic Reconstruction: Application to γ-Proteobacteria

Guillaume Blin[1], Cedric Chauve[2], and Guillaume Fertin[1]

[1] LINA FRE CNRS 2729, Université de Nantes,
2 rue de la Houssinière, BP 92208 - 44322 Nantes Cedex 3, France
{blin, fertin}@lina.univ-nantes.fr
[2] LaCIM et Département d'Informatique, Université du Québec À Montréal,
CP 8888, Succ. Centre-Ville, H3C 3P8, Montréal (QC), Canada
chauve@lacim.uqam.ca

Abstract. We study the problem of phylogenetic reconstruction based on gene order for whole genomes. We define three genomic distances between whole genomes represented by signed sequences, based on the matching of similar segments of genes and on the notions of breakpoints, conserved intervals and common intervals. We use these distances and distance based phylogenetic reconstruction methods to compute a phylogeny for a group of 12 complete genomes of γ-Proteobacteria.

Keywords: Phylogenetic reconstruction, breakpoints, common intervals, conserved intervals, γ-Proteobacteria, gene families.

1 Introduction

Methods based on gene orders have proved to be powerful for the study of evolution, both for eukaryotes [8,9] and for prokaryotes [11,2]. The main algorithmic methods developed for this purpose are based on a representation of a genome by a signed permutation (see several survey chapters in the recent book [12] for example). At first, this representation of genomes implies that these methods should be limited to the comparison of genomes having the exact same gene content and where there is a unique copy of each gene in each genome. This model thus fits perfectly with the study of gene order in mitochondrial genomes, for example [5]. However, in general, genomes do not share the same gene content or some gene families are not trivial – a given gene can occur more than once in a genome –, which implies that such genomes should be represented by signed sequences instead of signed permutations. There has been several attempts to develop methods for the comparison of such genomes and most of these methods are based on the transformation of the initial data, a set of signed sequences representing genomes, into a set of signed permutations, in order to apply one or several of the algorithms developed in this context. For example, the approach developed by the group of Pevzner for eukaryotic genomes is based on representing a genome by a sequence of *synteny blocks*, where such a block can contain several genes [8,9]. Another approach, developed by Sankoff [16], suppresses in

A. McLysaght et al. (Eds.): RECOMB 2005 Ws on Comparative Genomics, LNBI 3678, pp. 11–20, 2005.

every genome all but one copy of the genes of a gene family (the remaining gene of this family in a genome being called the *exemplar* gene), which leads to representing genomes by signed permutations. It is also natural to consider only a subset of the genes of a genome, that belong to families of size one, as it was done for a set of 30 γ-Proteobacteria in [2]. Finally, a recent approach is based on the computation of a matching of similar segments between two genomes that immediately allows to differentiate the multiple copies of a same gene and to represent genomes by signed permutations [19,6]. This method, combined with the *reversal distance between signed permutations*, has been shown to give good results on simulated data [19].

In the present work, we are interested in the computation of genomic distances between bacterial genomes, based on gene order for whole genomes, and to assess the quality of these distances for the reconstruction of phylogenies. We define three distances in terms of gene orders based on two main ingredients: (1) the computation of a matching of similar genes segments between two genomes, following the approach of [19], and (2) three measures of *conservation of the combinatorial structure*: *breakpoints, conserved intervals* and *common intervals*. This last aspect differs from most of previous works that relied on the reversal distance. Moreover, this is, as far as we know, the first time that distances based on conserved intervals and common intervals are used on real data. We test our distances on a set of 12 γ-Proteobacteria complete genomes studied in [15,11], and, for two different sets of gene families, we compute phylogenies for these data, using the Fitch-Margoliash method. We then compare the trees we obtain to the phylogenetic tree proposed in [15], based on a Neighbor-Joining analysis of the concatenation of 205 proteins.

2   Distances and Gene Matching

In this section, we introduce the combinatorial notions and algorithms used in the computation of distances based on gene order conservation for whole genomes.

Genomes Representation. We represent a genome by a *signed sequence* on the alphabet of *gene families*. Every element in a genome is called a *gene*[1] and belongs to a gene family. For a signed sequence G, one denotes by g_i the signed integer representing the ith gene in G. Two genes belong to the same gene family if they have the same absolute value.

Gene Matching. Given two signed sequences G and H, a *matching M* between G and H is a set of pairs (g_i, h_j), where g_i and h_j belong to the same gene family. Genes of G and H that do not belong to any pair of the matching M are said to be *unmatched* for M. A matching M between G and H is said to be *complete* if for any gene family, there are no two genes of this family that are unmatched for

---

[1] This terminology is restrictive as one could use the methods described in this work considering any kind of genetic marker located on a genome, but we follow the classical terminology and use the word gene through all this paper.

M and belong respectively to G and H. A matching M between G and H can be seen as a way to describe a putative assignment of orthologous pairs of genes between G and H (see [10] for example where this notion was used, together with the reversal distance, to the precise problem of orthologous assignment). In this view, segments of consecutive unmatched genes could represent segments of genes that have been inserted, by lateral transfer for example, or deleted, due to functional divergence of loss of genes after a lateral transfer or a segmental duplication for example, during the evolution.

Given a matching M between two genomes G and H, once the unmatched genes have been removed from these two genomes, the resulting matching M is a perfect matching between the remaining genes of the two genomes. It follows immediately that M defines a signed permutation of $|M|$ elements, denoted P_M, as illustrated in Figure 1. We also denote by $\mathrm{del}(G, M)$ and $\mathrm{del}(H, M)$ the number of maximum segments of consecutive unmatched genes in G and H.

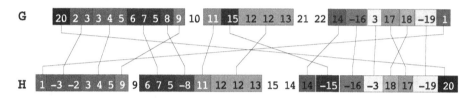

Fig. 1. A possible *complete matching* M between two genomes G and H represented as signed sequences. In this example, $\mathrm{del}(G, M) = \mathrm{del}(H, M) = 2$ and $P_M = 23$ -3 -2 4 5 6 11 7 8 9 -10 12 14 15 16 17 -13 18 -19 21 20 22 1.

Given G, H and a matching M between G and H, one can define a distance between G and H, induced by M, in terms of one of the classical distances based on signed permutations, applied to the permutation P_M, corrected with $\mathrm{del}(G, M)$ and $\mathrm{del}(H, M)$ in order to take into account modifications of gene order due to events like lateral transfer or loss of genes. In the following, we consider three different distances, based on three measures of the conservation of the combinatorial structure in signed permutations: breakpoints, conserved intervals and common intervals.

The rationale for using the above ideas in the design of a gene order distance between bacterial genomes relies on the observation that during their evolution, prokaryotic genomes seem to have been rearranged mostly by short reversals [14,18], which implies that close genomes will share similar clusters of genes [17]. Based on this hypothesis, one of the goals of our work was to study how distances based on the conservation of structure allow to capture phylogenetic signal, and we tried the three known measures of conservation of structures: breakpoints is the simplest and has been used for a long time, while the two other distances, based on intervals are more recent but capture more subtle similarities than breakpoints.

Breakpoints Distance. Let $P = p_1, \ldots, p_m$ be a signed permutation. A breakpoint in P is a pair of consecutive elements $p_i p_{i+1}$ such that $p_{i+1} \neq p_i + 1$. We denote by bkpts(P) the number of breakpoints of P. Given a matching M between G and H, and the corresponding signed permutation P_M, we define the breakpoints distance between G and H given M as follows:

$$d\_Breakpoints(G, H, M) = \frac{\text{bkpts}(P_M)}{|M|} + \frac{\text{del}(G, M)}{|G|} + \frac{\text{del}(H, M)}{|H|}$$

Note that this definition considers, in the computation of the distance, the size of the matching M and the size of the compared genomes, both characteristics that can vary a lot as it appears in our study of γ-Proteobacteria. In the example given in Figure 1, bkpts(P_M) = 14, and del(G, M) = del(H, M) = 2. We thus obtain $d\_Breakpoints(G, H, M) = \frac{14}{23} + \frac{2}{26} + \frac{2}{26} = 0,806$.

Distances Based on Intervals. The number of breakpoints in a signed permutation is a very natural measure of conservation of the structure of this permutation with respect to the identity permutation. Recently, several more complex measures of such structure conservation have been introduced, and in this work we consider two of them: conserved intervals and common intervals.

A common interval in a signed permutation P is a segment of consecutive elements of this permutation which, when one does not consider signs and order, is also a segment of consecutive elements of the identity permutation (see [3] for an example of the relationship between common intervals and the study of gene order). Conserved intervals of signed permutations were defined in [4]: a segment $p_i, \ldots, p_j$ of a signed permutation P, with $i \neq j$, is a conserved interval if it is a common interval of P and either $p_i > 0$ and $p_j = p_i + (j - i)$, or $p_i < 0$ and $p_j = p_i - (j - i)$ (in other words, in absolute value, p_i and p_j are the greatest and smallest elements of the common interval $p_i, \ldots, p_j$). For a given signed permutation P, one denotes respectively by ICommon(P) and IConserved(P), the number of common intervals in P and the number of conserved intervals in P.

Given a matching M between G and H, and the corresponding signed permutation P_M, we introduce here two new distances, based on ICommon(P_M) and IConserved(P_M) : one defines the common intervals distance between G and H given M by

$$d\_ICommon(G, H, M) = 1 - \frac{2 * \text{ICommon}(P_M)}{|M|^2} + \frac{\text{del}(G, M)}{|G|} + \frac{\text{del}(H, M)}{|H|}$$

and the conserved intervals distance between G and H given M by

$$d\_IConserved(G, H, M) = 1 - \frac{2 * \text{IConserved}(P_M)}{|M|^2} + \frac{\text{del}(G, M)}{|G|} + \frac{\text{del}(H, M)}{|H|}$$

Computation of a Matching. For a given distance model, a parsimonious approach for the comparison of two genomes G and H searches for a matching M between G and H involving the smallest distance between G and H. Unfortunately, this problem has been shown to be **NP**-complete, when using the breakpoints and conserved intervals distances [6,7]. Swenson *et al.* [19] proposed a fast heuristic to compute a matching based on a greedy approach consisting on (1) identifying the longest common segment of unmatched genes of G that is also a segment of unmatched genes in H, up to a reversal, (2) matching these two segments of G and H, and (3) repeating the process until a complete matching is found. In [7], Blin and Rizzi have designed a quite similar heuristic using a suffix-tree. We have used the heuristic of Swenson *et al.* in the present work. Let $M_{G,H}$ denote the matching returned by the heuristic with G as first and H as second parameters. As our implementation of the heuristic does not return a symmetric matching – matching $M_{G,H}$ may differ from $M_{H,G}$ –, we have defined the distances, respectively of breakpoints, conserved intervals and common intervals, between G and H as follows:

$$d\_Breakpoints(G, H) = \\ (d\_Breakpoints(G, H, M_{G,H}) + d\_Breakpoints(H, G, M_{H,G}))/2$$

$$d\_ICommon(G, H) = \\ (d\_ICommon(G, H, M_{G,H}) + d\_ICommon(H, G, M_{H,G}))/2$$

$$d\_IConserved(G, H) = \\ (d\_IConserved(G, H, M_{G,H}) + d\_IConserved(H, G, M_{H,G}))/2$$

3   Experimental Results and Discussion

Input Data. The data set we studied is composed of 12 complete genomes from the 13 γ-Proteobacteria studied in [15]. We have not considered the genome of *V.cholerae* because it is composed of two chromosomes, and this is not considered in our model. This data set is composed of the genomes of the following species: *Buchnera aphidicola APS* (Genbank accession number NC_002528), *Escherichia coli K12* (NC_000913), *Haemophilus influenzae Rd* (NC_000907), *Pasteurella multocida Pm70* (NC_002663), *Pseudomonas aeruginosa PA01* (NC_002516), *Salmonella typhimurium LT2* (NC_003197), *Xanthomonas axonopodis pv. citri 306* (NC_003919), *Xanthomonas campestris* (NC_0 03902), *Xylella fastidiosa 9a5c* (NC_002488), *Yersinia pestis CO_92* (NC_003143), *Yersinia pestis KIM5 P12* (NC_004088), *Wigglesworthia glossinidia brevipalpis* (NC_004344).

Data set and programs used and mentioned in this article can be found on a companion web site at `http://www.lacim.uqam.ca/~chauve/CG05`.

Gene Families. From these 12 genomes, the initial step was to compute a partition of the complete set of genes into gene families, where each family is supposed to represent a group of homologous genes. This partition induces the encoding of the genomes by signed sequences, that is the input of the matchings computation that leads to distance matrices. Hence, the result of a phylogenetic analysis based on gene order depends strongly on the initial definition of families. Due to this importance of the partition of genes into families, and in order to assess the quality of the distances we defined on our data set of γ-Proteobacteria genomes without relying on a single set of families, we used two different methods to partition genes into gene families. Both are based on alignments of amino-acid sequences with BLAST [1].

The first partition we used is the one computed in [15], in order to define families of orthologous genes used in a Neighbor-Joining analysis of these γ-Proteobacteria genomes, and has been provided to us by Lerat. Briefly, this partition is given by the connected components of a directed graph whose nodes are the coding genes and pseudo-genes of the 12 genomes and there is an edge from gene g to gene h if the bit-score of the BLAST comparison of g against h is at least equal to 30% of the bit-score of the BLAST comparison of g against itself. Details are available in [15].

To compute the second partition we used all coding genes of our 12 genomes, as well as ribosomal and transfer RNAs. For RNAs, the families were decided on the basis of the annotation of the genes. For coding genes, a family is a connected component of the undirected graph whose vertices are genes and where there is an edge between two genes g and h if the alignment computed by BLAST between the sequences of g and h has at least 25% of identity for both sequences, and overlaps at least 65% of both sequences.

We can notice that the matchings of the second partition are always bigger than the ones of the first partition. However, the difference between the two is always relatively small compared to the size of the matchings.

Details on partitions and matchings can be found on the companion web site.

Phylogenetic Trees Computation. Given a matrix distance, obtained by the algorithms described in Section 2, we computed phylogenetic trees using the following Fitch-Margoliash phylogenetic reconstruction method implemented in the `fitch` command (version 3.63) of the PHYLIP package available at `http://evolution. genetics.washington.edu/phylip.html`, where we have used the G (global rearrangements) and J (jumbling, with parameters 3 and 1000) options. We chose this method instead of the classical Neighbor-Joining method because it examines several tree topologies and optimizes a well defined criterion, based on the least-squared error. We have used the `retree` command of the PHYLIP package to re-root and flip some branches of the trees in order to harmonize the representation of our results with the tree obtained by Lerat *et al.* in [15–Figure 5].

Results and Analysis. Figures 2 and 3 present the trees obtained by applying our method on the breakpoints, common and conserved intervals distances, and the tree given by Lerat *et al.* using NJ method with the concatenation of 205 proteins [15–Figure 5], that we call the reference tree below.

One can notice that these trees agree relatively well with the reference tree. Indeed, we can stress the following points:

1. Using either set of gene families, one can notice that there are always differences that concern the taxa *Buchnera aphidicola* and *Wigglesworthia glossinidia brevipalpis*. However, Herbeck *et al.* [13] suggested that the fact that this clade exists in the results from Lerat *et al.* [15] is due to a bias in GC composition.

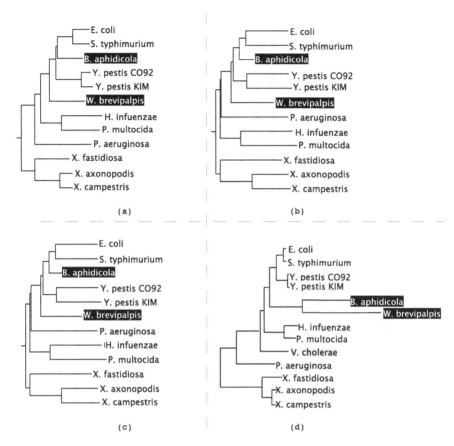

Fig. 2. Experimental results with the first set of gene families ([15]). (a) breakpoints distance. (b) common intervals distance. (c) conserved intervals distance. (d) reference tree obtained by Lerat *et al.* [15–Figure 5]. In gray, the genome not considered in our experiments. In black, *Buchnera aphidicola* and *Wigglesworthia glossinidia brevipalpis*.

2. Using the first partition, and if we do not consider the case of *Buchnera aphidicola* and *Wigglesworthia glossinidia brevipalpis* discussed above, one can notice that the tree obtained with the breakpoints distance agrees with the reference tree (Figure 2 (a)). Concerning the two other distances (conserved intervals and common intervals distances), the only difference lies in the position of *Pseudomonas aeruginosa* (Figures 2 (b) and 2 (c)).

3. Using the second partition, we also see that the tree obtained with the breakpoints distance agrees with the reference tree (Figure 3 (a)), if *Buchnera aphidicola* and *Wigglesworthia glossinidia brevipalpis* are not considered. Using any of the two other distances (conserved intervals and common intervals distances), the only difference concerns the group of taxa *Haemophilus influenzae* and *Pasteurella multocida*, that is placed at a different position (Figures 3 (b) and 3 (c)).

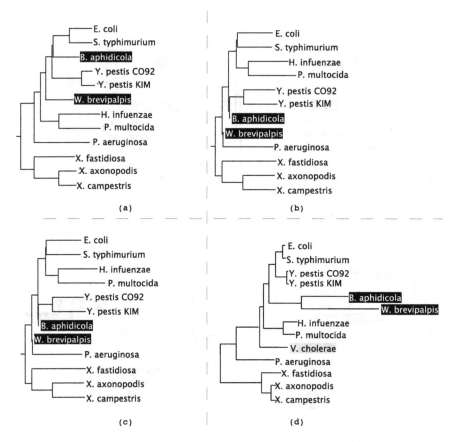

Fig. 3. Experimental results with the second set of gene families. (a) breakpoints distance. (b) common intervals distance. (c) conserved intervals distance. (d) reference tree obtained by Lerat *et al.* [15–Figure 5]. In gray, the genome not considered in our experiments. In black, *Buchnera aphidicola* and *Wigglesworthia glossinidia brevipalpis*.

Thus, we can say that the distances we defined capture a significant phylogenetic signal, and provide good results on real data. However, the use of distance relying on intervals, as opposed to the one based on breakpoints, seems to imply some inaccuracy in the trees we obtained. This should not come as a surprise, since our matching computation method is optimized for the breakpoints distance.

4   Conclusion

In this first study, we proposed a simple approach for the phylogenetic reconstruction for prokaryotic genomes based on the computation of gene matchings and distances expressed in terms of combinatorial structure conservation. Despite its simplicity, our approach gave interesting results on a set of 12 genomes of γ-Proteobacteria, as the trees we computed agree well with the tree computed in [15] and based on the concatenation of the sequences of 205 proteins. It should be noted that our results agree well too with another recent study based on gene order and signed permutations [2]. Moreover, this study raises several interesting questions.

First, the initial computation of gene families plays a central role in the gene order analysis. In [2] for example, where 30 γ-Proteobacteria genomes were considered, these families were designed in such a way that each one contains exactly one gene in every genome. As a consequence, if one considers all other genes as member of families of size one, there is only one possible matching for every pair of genomes. Based on these families, phylogenetic trees based on the reversal and breakpoints distances were computed. Our approach can be seen as less strict in the sense that pairwise genomes comparisons are not based only on genes that are present in all genomes, and our results agree quite well with the results of [2]. But more generally, it would be interesting to study more precisely the influence of the partition of the set of all genes into families on the whole process, and in particular the impact of the granularity of such a partition.

Second, a method for the validation of the computed trees, similar to the bootstrap commonly used in phylogenetic reconstruction, would be a very valuable tool. This lack of a validation step in our analysis was one of the main reasons that led us to use the Fitch-Margoliash method, that tries several topologies, instead of the Neighbor-Joining method. A validation method, based on a Jackknife principle was introduced in [2], but it was not clear how to use it in our context where the matchings used in pairwise comparisons can have very different sizes.

Finally, we think that an important point in the development of methods similar to the one described in this work should rely into the link between the computation of a matching and the kind of measure of structure conservation that is used to define a distance. Indeed, the principle of computing a matching by the identification of similar segments is natural when breakpoints are used, as two similar matched segments define only breakpoints at their extremities. But when using distances based on intervals, it would clearly be more interesting to consider also segments of similar gene content but maybe not with the same order of the genes.

References

1. S.F. Altschul, T.L. Maden, A.A. Schaffer, J. Zhang, Z. Zhang, W. Miller, and D.J. Lipman. Gapped blast and psi-blast: a new generation of protein database search programs. *Nucleic Acids Res.*, 25(17):3389–3402, 1997.
2. E. Belda, A. Moya, and F.J. Silva. Genome rearrangement distances and gene order phylogeny in γ-proteobacteria. *Mol. Biol. Evol.*, 22(6):1456–1467, 2005.
3. S. Bérard, A. Bergeron, and C. Chauve. Conserved structures in evolution scenarios. In *Comparative Genomics, RECOMB 2004 International Workshop*, volume 3388 of *Lecture Notes in Bioinformatics*, pages 1–15. Springer, 2004.
4. A. Bergeron and J. Stoye. On the similarity of sets of permutations and its applications to genome comparison. In *9th International Computing and Combinatorics Conference (COCOON '03)*, volume 2697 of *Lecture Notes in Computer Science*, pages 68–79. Springer, 2003.

5. M. Blanchette, T. Kunisawa, and D. Sankoff. Gene order breakpoint evidence in animal mitochondrial phylogeny. *J. Mol. Evol.*, 49(2):193–203, 1999.
6. G. Blin, C. Chauve, and G. Fertin. The breakpoints distance for signed sequences. In *1st International Conference on Algorithms and Computational Methods for Biochemical and Evolutionary Networks, CompBioNets 2004*, volume 3 of *Texts in Algorithms*, pages 3–16. KCL Publications, 2004.
7. G. Blin and R. Rizzi. Conserved interval distance computation between non-trivial genomes. In *11th International Computing and Combinatorics Conference (CO-COON '05)*, 2005. To appear in *Lecture Notes in Computer Science*.
8. G. Bourque, P.A. Pevzner, and G. Tesler. Reconstructing the genomic architecture of ancestral mammals: lessons from human, mouse and rat genomes. *Genome Res.*, 14(4):507–516, 2004.
9. G. Bourque, E.M. Zdobnov, P. Bork, P.A. Pevzner, and G. Tesler. Comparative architectures of mammalian and chicken genomes reveal highly variable rates of genomic rearrangements across different lineages. *Genome Res.*, 15(1):98–110, 2005.
10. X. Chen, J. Zheng, Z. Fu, P. nan, Y. Zhong, S. Lonardi, and T. Jiang. Computing the assignment of orthologous genes via genome rearrangement. In *3rd Asia-Pacific Bioinformatics Conference 2005*, pages 363–378. Imperial College Press, 2005.
11. J.V. Earnest-DeYoung, E. Lerat, and B.M.E. Moret. Reversing gene erosion: Reconstructing ancestral bacterial genomes from gene-content and order data. In *Algorithms in Bioinformatics, 4th International Workshop, WABI 2004*, volume 3240 of *Lecture Notes in Bioinformatics*, pages 1–13. Springer, 2004.
12. O. Gascuel, editor. *Mathematics of Evolution and Phylogeny*. Oxford University Press, 2005.
13. J.T. Herbeck, P.H. Degnan, and J.J. Wernegreen. Nonhomogeneous model of sequence evolution indicates independent origins of primary endosymbionts within the enterobacteriales (γ-proteobacteria). *Mol. Biol. Evol.*, 22(3):520–532, 2004.
14. J-F. Lefebvre, N. El-Mabrouk, E. Tillier, and D. Sankoff. Detection and validation of single gene inversions. *Bioinformatics*, 19(Suppl. 1):i190–i196, 2003.
15. E. Lerat, V. Daubin, and N.A. Moran. From gene tree to organismal phylogeny in prokaryotes: the case of γ-proteobacteria. *PLoS Biology*, 1(1):101–109, 2003.
16. D. Sankoff. Genome rearrangement with gene families. *Bioinformatics*, 15(11):909–917, 1999.
17. D. Sankoff. Short inversions and conserved gene clusters. *Bioinformatics*, 18(10):1305–1308, 2002.
18. D. Sankoff, J-F. Lefebvre, E. Tillier, A. Maler, and N. El-Mabrouk. The distribution of inversion lengths in prokaryotes. In *Comparative Genomics, RECOMB 2004 International Workshop, RCG 2004*, volume 3388 of *Lecture Notes in Bioinformatics*, pages 97–108. Springer, 2005.
19. K.M. Swenson, M. Marron, J.V Earnest-DeYoung, and B.M.E. Moret. Approximating the true evolutionary distance between two genomes. In *Proceedings of the seventh Workshop on Algorithms Engineering and Experiments and Second Workshop on Analytic Algorithmics and Combinatorics (ALENEX/ANALCO 2005)*. SIAM, 2005.

Maximizing Synteny Blocks to Identify Ancestral Homologs

Guillaume Bourque[1], Yasmine Yacef[2], and Nadia El-Mabrouk[2]

[1] Genome Institute of Singapore, 138672, Singapore
bourque@gis.a-star.edu.sg
[2] DIRO, Université de Montréal, H3C 3J7, Canada
mabrouk@iro.umontreal.ca

Abstract. Most genome rearrangement studies are based on the assumption that the compared genomes contain unique gene copies. This is clearly unsuitable for species with duplicated genes or when local alignment tools provide many ambiguous hits for the same gene. In this paper, we compare different measures of order conservation to select, among a gene family, the pair of copies in two genomes that best reflects the common ancestor. Specifically, we present algorithms to identify ancestral homologs, or exemplars [1], by maximizing synteny blocks between genomes. Using simulated data, we validate our approach and show the merits of using a conservative approach when making such assignments.

1 Introduction

Identifying homologous regions between genomes is important, not only for genome annotation and the discovery of new functional regions, but also for the study of evolutionary relationships between species. Once orthologous genes have been identified, the genome rearrangement approach infers divergence history in terms of global mutations, involving the displacement of chromosomal segments of various sizes. The major focus has been to infer the most economical scenario of elementary operations transforming one linear order of genes into another. In this context, inversion (or "reversal") has been the most studied rearrangement event [2–6], followed by transpositions [7–9] and translocations [10–12]. All these studies are based on the assumption that each gene appears exactly once in each genome, which is clearly an oversimplification for divergent species containing paralogous and orthologous gene copies scattered across the genome. Moreover, even for small genomes (viruses, bacteria, organelles) where the hypothesis of no paralogy may be appropriate, the assumption of a one to one correspondence between genes assumes a perfect annotation step. However, in many cases, the similarity scores given by the local alignment tools (such as BLAST or FASTA) are too ambiguous to conclude to a one to one homology, and using different parameters and cut-off values may lead to different sets of orthologs.

Approaches to identify homology typically only rely on local mutations; they neglect the genomic context of each gene copy that might provide additional information. For example, if two chromosomes are represented by the two gene orders "*badc*" and "*badceaf*", the two first *a*'s are more likely to be the two

A. McLysaght et al. (Eds.): RECOMB 2005 Ws on Comparative Genomics, LNBI 3678, pp. 21–34, 2005.

copies derived from the common ancestor, as they are preserving the gene order context in the two chromosomes. Sankoff [1] was the first to test this idea with the *exemplar approach*. The underlying hypothesis is that in a set of homologs, there commonly exists a gene that best reflects the original position of the gene family ancestor. The basic concept of Sankoff's algorithm is to remove all but one member of each gene family in each of the two genomes being compared, so as to minimize the breakpoint or the reversal distance. Context conservation has also been used in the annotation of bacterial genomes [13] to choose, among a set of BLASTP best hits, the true ancestral copies, also called *positional homologs*. We now want to extend these ideas to other measures of gene order conservation such as conserved and common intervals [14–17]. These alternatives measures generalize the breakpoint distance and similarly allow the comparison of a set of genomes. Moreover, they allow the study of global genome evolution without focusing on a specific rearrangement model.

In this paper, we use the common and conserved interval criteria to identify the ancestral homologs. Generalizing the fact that gene copies that are surrounded by the same genes in different genomes are more likely to be the true ancestral copies, we identify ancestral homologs by maximizing blocks of synteny between genomes. In Section 2, we review some gene order measures and their use for genome rearrangement with gene families. In Section 3, we describe our method and present algorithms for ancestral homolog assignment. In Section 4, we analyze the performance of our method using simulated data and show the effect of homolog assignment on the induced rearrangement distance.

2   Related Work

In the rest of this paper, a gene family a will refer to all homologs (orthologs and paralogs) of a gene a among a set of genomes. Orthologs are copies among different genomes that have evolved by speciation while paralogs are copies that have evolved by duplication. A genome will be considered uni-chromosomal and represented as a linear order of signed genes, where the sign represents the transcriptional orientation of the gene. A chromosomal *segment* $[a, b]$ is just the subsequence surrounded by the two genes a and b.

2.1   Genome Rearrangement with Gene Families

Gene orders can be compared according to a variety of criteria. The *breakpoint distance* between two genomes G and H measures the number of pairs of genes a, b that are adjacent in one genome (contains the segment '$a\ b$') but not in the other (contains neither '$a\ b$' nor '$-b\ -a$'). *Rearrangement distances* measure the minimal number of genome rearrangements (inversions, transpositions, translocations$\cdots$) necessary to transform one order of genes into another.

Most work on rearrangement has been restricted to the comparison of genomes with no gene copies. A method that does takes into account duplications, but requires that the number of copies is the same in both genomes, has been presented by Tang and Moret [18]. Their approach relied on a straightforward

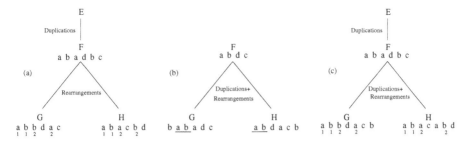

Fig. 1. (a) Evolutionary model considered in [19]; using the breakpoint distance, the chosen homologs are the one underlined by the same number in both genomes. (b) Model considered in [1]; using the breakpoint distance, the chosen exemplars are the underlined ones. (c) General model with duplications occurring before and after speciation; using the breakpoint distance and running the exemplar algorithm twice, the chosen homologs are the one underlined by the same number in both genomes.

enumeration of all possible assignments of homologs between two genomes. More recently, Chen *et al.* [19] gave an NP-hard result for this problem under the reversal distance and presented an efficient heuristic based on a maximal cycle decomposition of the Hannenhalli and Pevzner breakpoint graph [10,3]. Both of these studies are based on an evolutionary model assuming that all copies were present in the common ancestor and no duplication occurred after speciation (Fig. 1a). In many contexts, this assumption may be questionable.

Another approach relaxing the copy number constraint has been considered by Sankoff [1]. The *exemplar approach* consists in deleting, from each gene family, all copies except one in each of the compared genomes G and H, so that the two resulting permutations have the minimal breakpoint or reversal distance. The underlying evolutionary model is that the most recent common ancestor F of genomes G and H has single gene copies (Fig. 1b). After divergence, the gene a in F can be duplicated many times in the two lineages leading to G and H, and appear anywhere in the genomes. Each genome is then subject to rearrangement events. After rearrangements, the direct descendent of a in G and H will have been displaced less frequently than the other gene copies. Even though finding the positional homologs (called *exemplars* in [1]) has been shown NP-hard [20], Sankoff [1] developed a branch-and-bound algorithm that has been shown practical enough on simulated data. More recently, Nguyen *et al.* [21] developed a more efficient divide-and-conquer approach.

The preceding model is based on the hypothesis of a unique ancestral copy for each gene family. However, in the more general case of an ancestral genome containing paralogs, for each gene family, not only one but many pairs of ancestral homologs have to be found (Fig. 1c). The exemplar approach can also be applied to this model. Indeed, by running the algorithm n times, n homolog assignments are made for the same gene family. Recently, Blin *et. al* [22] gave an NP-hard result and proposed a branch-and-bound exact algorithm to compute the breakpoint distance under this model.

Finally, Marron *et al.* [23] and Swenson *et al.* [24] have also developed an alternative framework for the study of the evolution of whole genomes with unequal gene content. In this approach, instead of computing an edit distance, the evolutionary distance is estimated by analyzing a different measure of conservation called the minimum cover. In combination with a neighbor-joining procedure, this method was shown to successfully recover simulated trees under various conditions [24].

2.2 Synteny Blocks

The drawback of considering a rearrangement distance to compare genomes is the strong underlying model assuming evolution by one or two specific rearrangement events. A simpler measure of order conservation (synteny) is the breakpoint distance. Other more general measures of synteny have been proposed in the genome rearrangement literature [14,16,17] and are now being reviewed.

Conserved Blocks. The notion of *conserved intervals* or *blocks* that has been introduced in [14] is identical to the notion of a *subpermutation* introduced in the Hannenhalli and Pevzner theory [10]. It is defined for genomes with single gene copies as follows.

Definition 1. *Given two genomes G and H, a conserved block is defined by two signed genes a and b and a set of unsigned genes U such that, in each genome, there exists a segment of the form $S = [a, b]$ or $S = [-b, -a]$, and the set of unsigned genes appearing between the two endpoints is U (Fig. 2a). Such a conserved block will be denoted $[a, U, b]$.*

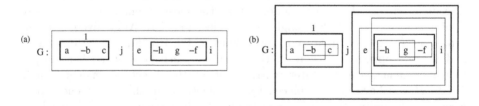

Fig. 2. The blocks of G and H, for H being the identity permutation $abcdefghij$. (a) Rectangles represent conserved blocks. For example, rectangle 1 represents the block $[a, U, c]$ with $U = \{b\}$. Bold rectangles are minimal blocks (not containing any other block); (b) Common blocks. For example, rectangle 1 represents the common block $\{a, b, c\}$. Bold rectangles are commuting blocks (either contained or have an empty intersection with any other block).

For genomes with gene copies, the problem of finding a pairing of gene copies that maximizes the number of conserved blocks (i.e. minimizing the conserved block distance) has been recently shown to be NP-complete [15].

Common Blocks. Even though conserved blocks have been shown useful for the genome rearrangement studies, the endpoint constraint contained in the definition is not directly linked to a specific biological mechanism. The notion of a common block introduced in [16] relaxes this constraint.

Definition 2. *Let G and H be two genomes on the gene set $\{c_1, \cdots, c_n\}$. A subset C of $\{c_1, \cdots, c_n\}$ is a* common block *of G and H iff G (respec. H) has a segment which unsigned gene content is exactly C.*

Common blocks have been considered as an additional criteria to improve the realism of genome rearrangement scenarios [17,16].

3   Maximizing the Blocks

Following the assumption that the true descendents of an ancestral gene in two genomes are the copies that have been less rearranged, the objective is to find a pairing of gene copies that maximizes gene order conservation. We use two measures of order conservation: the total number of conserved blocks and the total number of common blocks.

There are a number of reasons to maximize the number of synteny blocks. First, the more blocks we can construct among a set of genomes, the farther they are from random permutations. Indeed, random orders are less likely to share large intervals of similar content. Second, they generalize the breakpoint criteria used in previous ancestral homolog assignment methods [1,18,19,13]. Third, in contrast with rearrangement distances, they allow to model and compare, not only two genomes, but also a set of genomes. Finally, although conserved blocks are not directly linked to a specific rearrangement event, they represent the components of the Hannenhalli and Pevzner graph [3,10], and as such, are related to reversals.

It is preferable to measure similarity using the total number of blocks instead of the number of minimal or commuting blocks mostly because two overlapping blocks denote a better conservation than two disjoint blocks. Taking minimal blocks or commuting blocks alone does not reflect this difference. In contrast, maximizing the total number of blocks creates a bias towards overlapping blocks and tend to favor small local rearrangements, which is justified by a variety of biological and theoretical studies [25,26].

In section 3.1, we adapt the concepts of common and conserved blocks for genomes with gene families. Next, we describe the main steps of our algorithm. The first step, described in section 3.2, involves identifying all putative conserved blocks in a genome. This is accomplished by building a tree-like structure to store the conditions under which every possible block is attainable. In section 3.3, we show how intersecting these trees allows to efficiently identify the conditions required for these blocks to be conserved across multiple genomes. Finally, in section 3.4, we describe how to select the gene copies in order to maximize the total number of conserved blocks.

3.1   Blocks for Genomes with Gene Families

A *homolog assignment* is a procedure that connects, from each gene family, two particular gene copies, one from each genome. We generalize the notion of blocks (conserved or common) to two sequences containing gene copies as follows.

Definition 3. *Let G be a genome on the gene family set $\{c_1, \cdots, c_n\}$. An individual common block of G is any subset C of $\{c_1, \cdots, c_n\}$ that can be obtained from any segment of G and any homolog assignment. An individual conserved block is similar but is defined by its endpoints (c_i, c_j) and the gene subset U contained between these endpoints. Given two genomes G and H with possible gene copies, a* common *(respec.* conserved*) block is an individual common (respec. conserved) block of both G and H.*

For example, $\{a, b, c, f\}$ is an individual common block of the genome G in Fig. 3 obtained by choosing the copy f_1 from the gene family f. It is also an individual common block of H obtained by choosing the copy d_2 in the gene family d. Therefore, it is a common block of G and H. On the other hand, G contains two individual conserved blocks ending with a and c, depending on whether f_1 is the copy chosen from the gene family f, or not (Fig. 3.(1)). In the former case the block is B_1, ending with a, c and defined by $U = \{b, f\}$; in the latter case, the block is B_2 ending with a, c and defined by $U = \{b\}$. B_1 is a conserved block of G and H, as it is also an individual conserved block of H (by choosing c_2, d_2 and any of the two copies b_1 or b_2) (Fig. 3.(2)). B_2 is also a conserved block of G and H, as it is an individual block of H (by choosing b_1 and c_1). However blocks B_1 and B_2 are incompatible in H as they require two different homolog assignments for the gene family c (Fig. 3.(3)).

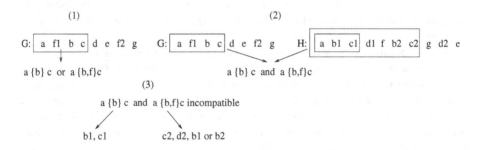

Fig. 3. Maximizing the number of conserved blocks for the genomes G and H with seven gene families represented by a, b, c, d, e, f, g. (1) Finding individual conserved blocks in G; (2) Finding conserved blocks; (3) Maximizing the compatible conserved blocks.

Our method will consist of three steps: 1) find all individual blocks of G and H respectively, 2) find the common or conserved blocks by superimposing the individual blocks of G and H and 3) select a maximal number of compatible conserved or common blocks. The method used at steps 2 and 3 is identical for common and conserved blocks. However, step 1 is slightly different for the two criteria. We will present the method for conserved blocks, and indicate the differences for common blocks.

3.2   Finding Individual Conserved Blocks

For each genome and each pair $\{a, b\}$ representing two gene families, we compute all individual conserved blocks $[a, U, b]$ by traversing the genome once, and constructing a tree-like structure $\mathcal{T}_{a,b}$ (Fig. 4a,b). The initial node is denoted by Φ. At the end of the construction, a *terminal node* t represents an individual block $[a, U, b]$ defined by the set U of labels in the path from Φ to t. As a block is not affected by the order of its elements between its two endpoints, for efficiency purposes we maintain a lexicographical order for each path in the tree.

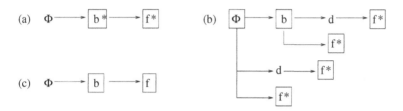

Fig. 4. The trees obtained for the pair $\{a, c\}$ for (a): genome G, and (b): genome H, first introduced in Fig. 3. Marked states are denoted by a '*', and terminal states are boxed. Superimposing trees (a) and (b) gives the tree (c), which represents the common blocks $[a, c]$ of G and H: the one containing b, and the one containing $\{b, f\}$.

During the tree construction, in addition to be terminal or not, each node is either marked or unmarked. The marked nodes correspond to partial individual blocks that can potentially form individual blocks later if a gene b is encountered.

At the beginning, $\mathcal{T}_{a,b}$ is restricted to a marked initial node Φ. The segment surrounded by the first copy of a and the last copy of b is then traversed from left to right. For each gene S_i in this segment, if $S_i = a$, we mark the initial state; if $S_i = b$, all marked states become terminal; otherwise, the tree $\mathcal{T}_{a,b}$ is incremented by *Algorithm Add-Node* (Fig. 5). For simplicity, we do not distinguish between a node and its label. Moreover, the *lexicographical order* of *node* refers to the lexicographical order of the sequence of labels in the path from Φ to *node*.

Finally, a terminal path denotes a path from Φ to a terminal node. A nonterminal path denotes any path from Φ or a terminal node to a leaf, that does not contain any terminal node.

Theorem 1. $[a, U, b]$ *is an individual conserved block of* G *if and only if* U *is the set of node's labels of a unique terminal path of* $\mathcal{T}_{a,b}$.

Algorithm Add-Node $(S_i,\ \mathcal{T}_{a,b})$

1. **For** each *node* in lexicographical order **Do**
2.   **If** *node* $=\ S_i$
3.   Mark *node*;
4.   **Else If** *node* $<\ S_i$ **and** *node* is marked
5.   **If** *node* has a child labeled S_i
6.   Mark this child;
7.   **Else**
8.   Create a node *new* labeled S_i, and an edge from *node* to *new*;
9.   Mark *new*;
10.   **End If**
11.   **Else If** *node* $>\ S_i$
12.   *nodePrec* $=$ *node*'s father; $P =$ subtree rooted by *nodePrec*;
13.   **If** *nodePrec* does not have a child labeled S_i
14.   Create node *new* labeled S_i, and an edge from *nodePrec* to *new*;
15.   **End If**
16.   Attach P to the child of *nodePrec* labeled S_i;
17.   **End If**
18.   **If** *node* $\geq\ S_i$
19.   Skip all nodes of the subtree rooted by *node*
20.   **End If**
21. **End For**
22. **If** S_i represents a single gene
23.   Remove all non-terminal paths that do not contain S_i;
24.   Unmark the nodes in all terminal paths that do not contain S_i;
25.   Unmark all the ancestors of the S_i nodes;
26. **End If**

Fig. 5. Updating the tree $\mathcal{T}_{a,b}$ after reading the next gene S_i in the largest segment of genome G surrounded by the gene families a and b

Complexity. For each of the n^2 gene pairs $\{a,b\}$, where n is the number of genes, each genome G and H is traversed once. For each pair $\{a,b\}$ and each position i (from 1 to the size m of the genome), the ith character G_i of G has to be added to the current tree $\mathcal{T}_{a,b}$. This requires the traversal of the tree once, and potentially perform subtree copies. Therefore, the worst time complexity of the algorithm is in $O(2n^2mS)$, where S is the size of the largest tree. In practice, subtree copies can be time consuming for large trees, making the algorithm inapplicable for large data. But, an easy way to circumvent this problem is to fix a tree depth threshold limiting the search to blocks of bounded length. We will show in Section 4.1 that using any reasonable tree depth threshold provides similar levels of accuracy.

Common Blocks. In the case of common blocks, there are three main differences: 1) we construct a unique tree for each genome (instead of constructing a tree for each pair $\{a, b\}$ and each genome), 2) the initial state Φ is always marked and 3) all tree-states are terminal.

3.3   Finding All Conserved Blocks

The conserved blocks $[a, U, b]$ of G and H are obtained by superimposing the two trees $T_{a,b}^G$ and $T_{a,b}^H$ corresponding to G and H respectively (Fig. 4c).

Theorem 2. $[a, U, b]$ *is a common block of* G *and* H *if and only if* U *is the set of node's labels of a terminal path common to* $T_{a,b}^G$ *and* $T_{a,b}^H$.

Notice that not all gene families are contained in conserved blocks. Consequently some gene families that have not retain sufficient positional context in both genomes may not be "resolved" with our approach. For example, the tree of Fig. 4c does not contain nodes for gene families d and c.

3.4   Maximizing Compatible Blocks

As illustrated in Fig. 3c, different blocks are obtained by different constraints that may be contradictory. In order to find compatible blocks, the constraints attached to each block have to be computed during the construction of individual trees. This is done with no additional complexity cost, by just labeling node marks, and keeping in a table all constraints attached to each mark. As soon as an endpoint is encountered, all marks become terminal and are reported with their constraints (Fig. 6).

Finally, after superimposing the two genome's trees and amalgamating the corresponding constraints at each terminal state, a set of clauses representing

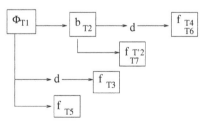

	b1	b2	c1	c2	d1	d2
T1	0	1	1	0		
T2	1	0	1	0		
T'2			0	1	0	1
T3			0	1	1	0
T4			0	1	1	0
T5	0	0	0	1	0	1
T6	0	1	0	1	1	0
T7	0	1	0	1	0	1

Fig. 6. Terminal states of the tree in Fig. 4b. The table represents states constraints: 1s are variables that have to be chosen, and 0s those that have to be avoided. Empty squares mean no constraints for the corresponding variables. State T_5 is irrelevant and has to be removed in a subsequent step, as b_1 and b_2 can not be avoided simultaneously.

all conserved blocks is obtained. Maximizing the number of compatible blocks is then reduced to a problem related to the extensively studied maximum satisfiability (MAX-SAT). It is stated as follows: given a boolean formula in conjunctive normal form (CNF), find a truth assignment satisfying a maximum number of its clauses. Even though the MAX-SAT problem is NP-complete, it is well characterized, and many efficient heuristics have been developed. However, the clauses representing our blocks are not in CNF. Therefore, no direct MAX-SAT solver can be used in this case. We developed an appropriate heuristic based on the general method classically used to solve MAX-SAT problems: 1) Set an initial solution (variable assignment) and evaluate the clauses; 2) Explore a neighborhood of the initial solution, reevaluate the clauses and keep the best solution; 3) Stop at convergence or after a fixed number of iterations.

4   Experimental Results

We used simulated data to assess the performance of the synteny blocks criteria to assign ancestral homologs. The data is generated as follows. Starting from a genome G with 100 distinct symbols representing 100 gene families, we obtain a second genome H by performing k rearrangements on G, and then randomly adding p_G gene copies in G and p_H gene copies in H at random positions. These copies may represent artifacts of an alignment tool. We simulated 5 different instances for each triplet (k, p_G, p_H), for $k \in \{10, 30, 50, 70, 90\}$ and $(p_G, p_H) \in \{(0, 10), (0, 20), (10, 10)\}$. We considered two rearrangement models: 1) inversions, transposition and inverted transpositions of size l following a Poisson distribution $P_\lambda(l)$ with $\lambda = 0.8$, to favor rearrangements of short segments (ALL) and 2) inversions of random size only (INV). We then run the algorithms and consider the number of correct homolog assignments (resolved) and false predictions.

4.1   Impact of Tree Depth Threshold

As explained in Section 3.2, in order to obtain an efficient time algorithm, we use a heuristic that constructs individual trees not exceeding a given tree depth threshold. Fig. 7 shows the results obtained for tree depth thresholds 5 and 10, using the evolutionary model ALL. For both common and conserved blocks, there is very little tree depth effect on the quality of the result. In general, depth 10 does provides slightly better results with a few more resolved genes and slightly fewer false predictions. This result validates the fact that restricting the search to blocks of limited size is sufficient to capture the genomic context information. But the extent to which this is true would need to be tested further by increasing, for instance, the level of duplications.

4.2   Comparing Synteny Blocks and Breakpoint Distance Criteria

We have compared the blocks criteria with the breakpoint distance criteria using the exemplar method developed in [21]. Fig. 8a shows the results obtained for the evolutionary model ALL. In general, the conserved and common blocks criteria

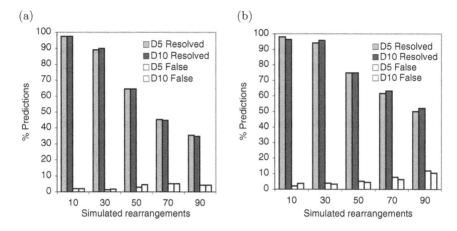

Fig. 7. Tree depth effect, for depth 5 (D5) and depth 10 (D10), on homolog assignment using: (a) the conserved block and (b) the common block criterion. Simulated genomes have 100 gene families. For a given number of rearrangements k under the evolutionary model ALL, the results of 15 instances, five for each $(p_G, p_H) \in \{(0, 10), (0, 20), (10, 10)\}$, are averaged.

allow to correctly resolve fewer genes than the exemplar method. However, the number of false predictions is notably reduced with our approaches. Comparing the two blocks criteria, common blocks correctly resolve more genes, while conserved blocks give fewer false predictions. We further compared the common and

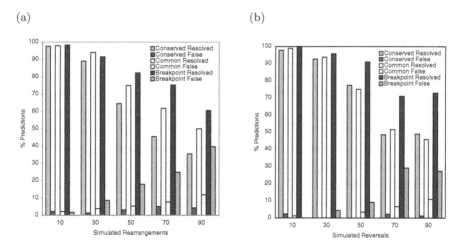

Fig. 8. (a) Comparison of the blocks and breakpoint distance criteria, using evolutionary model ALL; (b) Comparison of the blocks criteria using evolutionary model INV. For a given number of rearrangements k, the results of 15 instances, five for each $(p_G, p_H) \in \{(0, 10), (0, 20), (10, 10)\}$, are averaged.

conserved blocks criteria using the evolutionary model INV (Fig. 8b). It appears that both criteria have almost the same proportion of true predictions, while the common blocks criterion produces more false predictions. The advantage of the conserved block criteria under this model might be related to the its link to the reversal theory [14].

4.3   Impact of Homolog Assignment on the Reversal Distance

Various approaches have been considered in the past to preprocess duplicated genes for genome rearrangement studies. A common approach has been to remove all duplicated genes even though the missing data will typically lead to an underestimate of the rearrangement distances. An alternative approach could be to randomly assign corresponding pairs but that, in contrast, would lead to an overestimate of the actual distances. We were interested in measuring the extent of this under/over estimation and to compare it with the bias of our own methods for homolog assignment. The results are shown in Fig. 9.

We observe a much stronger impact, especially at moderate levels of rearrangements, of the random assignment of homologs compared to the simple removal of all duplicated genes. When $k < .4n$, apart from *random*, all methods are virtually indistinguishable making *none* a very acceptable method to estimate the rearrangement distance (at least at this level of duplications).

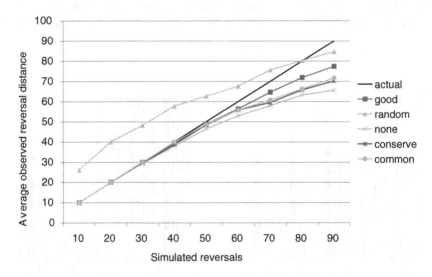

Fig. 9. The impact of homolog assignment on the observed reversal distance. 'Actual': number of simulated reversals. 'Good': ideal case with true ancestral assignment, 'Random': random selection of orthologs. 'Conserve' and 'common' are as before. For a given number of reversals k, the results of 5 instances for $(p_G, p_H) = (0, 20)$ are averaged. Reversal distances between the final permutations are computed using GRAPPA [27].

5   Conclusion

We have shown how synteny blocks can be used to accurately recover a large proportion of ancestral homologs. The same approach could directly be used to improve basic orthology prediction since it would not only rely on local alignment but also on genomic context. But, the development of a more hybrid approach that would combine both measures dynamically would also be desirable.

Based on the observation that incorrect assignment of homologs tend to have a more damageable impact on the induced rearrangement distances, we propose that a conservative approach, with a low level of false positives, is probably most desirable for this problem. Another strength of the approach is that it can directly be generalized to sets of multiple genomes. The next step of our work will be use the method, as an alternative to the one used in [13], to assign positional homology in bacterial genomes, and subsequently for the annotation of more complex genomes.

Acknowledgments. We are grateful to Louxin Zhang for his participation in the experimental part of the project. We are also thankful to Elisabeth Tillier for fruitful discussions and to the anonymous reviewers for thoughtful suggestions. GB is supported by funds from the Agency for Science, Technology and Research (A*STAR), Singapore.

References

1. D. Sankoff. Genome rear. with gene fam. *Bioinformatics*, 15:909–917, 1999.
2. J. Kececioglu and D. Sankoff. Exact and approx. algo. for sorting by reversals, with application to genome rear. *Algorithmica*, 13:180–210, 1995.
3. S. Hannenhalli and P. A. Pevzner. Transforming cabbage into turnip (polynomial algorithm for sorting signed permutations by reversals). *J. ACM*, 48:1–27, 1999.
4. H. Kaplan, R. Shamir, and R. E. Tarjan. A faster and simpler algorithm for sorting signed permutations by reversals. *SIAM Journal on Computing*, 29:880–892, 2000.
5. N. El-Mabrouk. Genome rearrangement by reversals and insertions/deletions of contiguous segments. In *CPM 2000*, volume 1848 of *LNCS*, pages 222- 234, 2000.
6. A. Bergeron. A very elementary presentation of the Hannenhalli-Pevzner theory. In *CPM*, LNCS. Springer Verlag, 2001.
7. Vineet Bafna and Pavel A. Pevzner. Sorting by transpositions. *SIAM Journal on Discrete Mathematics*, 11(2):224–240, 1998.
8. M. E. Walter, Z. Dias, and J. Meidanis. Reversal and transposition distance of linear chromosomes. In *String Proc. Information Retrieval (SPIRE '98)*, 1998.
9. T. Hartman. A simpler 1.5-approximation algorithm for sorting by transpositions. In *LNCS 2676*, volume CPM'03, pages 156-169, 2003.
10. S. Hannenhalli and P.A. Pevzner. Transforming men into mice (polynomial algorithm for genomic distance problem). In *Proc. IEEE 36th Ann. Symp. Found. Comp. Sci.*, pages 581–592, 1995.
11. G. Tesler. Efficient algorithms for multichromosomal genome rearrangements. *J. Comp. System Sci.*, 65(3):587-609, 2002.
12. M. Ozery-Flato and R. Shamir. Two notes on genome rearrangnements. *J. of Bioinf. and Comput. Biol.*, 1(1):71-94, 2003.

13. I.J. Burgetz, S. Shariff, A. Pang, and E. Tillier. Positional homology in bacterial genomes. manuscript, 2005.
14. A. Bergeron and J. Stoye. On the similarity of sets of permutations and its app. to genome comparison. In *COCOON'03*, volume 2697 of *LNCS*, pages 68- 79, 2003.
15. G. Blin and R. Rizzi. Conserved interval distance computation between non-trivial genomes. In *COCOON'05*, 2005.
16. M. Figeac and J.S. Varré. Sorting by reversals with common intervals. In *LNBI*, volume 3240 of *WABI 2004*, pages 26 - 37. Springer-Verlag, 2004.
17. S. Bérard, A. Bergeron, and C. Chauve. Conservation of combinatorial structures in evolution scenarios. In *LNCS*, volume 3388 of *RECOMB 2004 sat-meeting comp. gen.*, pages 1 - 14. Springer, 2004.
18. J. Tang and B.M.E. Moret. Phylogenetic reconstruction from gene rearrangement data with unequal gene contents. In *8th Workshop Algo. Data Struct. (WADS'03)*, volume 2748 of *LNCS*, pages 37-46. Springer Verlag, 2003.
19. X. Chen, J. Zheng, Z. Fu, P. Nan, Y. Zhing, S. Lonardi, and T. Jiang. Assignment of orthologous genes via genome rearrangement. *TCCB*, 2005. accepted.
20. D. Bryant. The complexity of calculating exemplar distances. In *Comparative Genomics: Empirical and Analytical Approaches to Gene Order Dynamics, Map alignment and the Evolution of Gene Families*, volume 1 of *Series in Computational Biology*. Kluwer Academic Press, 2000.
21. C.T. Nguyen, Y.C. Tay, and L. Zhang. Divide-and-conquer approach for the exemplar breakpoint distance. *Bioinformatics*, 2005.
22. G. Blin, C. Chauve, and G. Fertin. The breakpoint distance for signed sequences. In *Texts in Algorithm*, volume 3, pages 3- 16. KCL publications, 2004.
23. M. Marron, K.M. Swenson, and B.M.E. Moret. Genomic distances under deletions and insertions. *Theoretical Computer Science*, 325, 3:347–360, 2004.
24. K.M. Swenson, M. Marron, J.V. Earnest-DeYoung, and B.M.E. Moret. Approximating the true evolutionary distance between two genomes. In *7th Workshop on Algorithm Engineering and Experiments (ALENEX'05)*. SIAM Press, 2005.
25. J.F. Lefebvre, N. El-Mabrouk, E. Tillier, and D. Sankoff. Detection and validation of single gene inversions. *Bioinformatics*, 19:190i-196i, 2003.
26. P. Pevzner and G. Tesler. Human and mouse genomic sequences reveal extensive breakpoint reuse in mamm. evol. *PNAS, U.S.A.*, 100(13):7672-7677, 2003.
27. http://www.cs.unm.edu/~moret/GRAPPA.

An Expectation-Maximization Algorithm for Analysis of Evolution of Exon-Intron Structure of Eukaryotic Genes

Liran Carmel, Igor B. Rogozin, Yuri I. Wolf, and Eugene V. Koonin

National Center for Biotechnology Information,
National Library of Medicine,
National Institutes of Health,
Bethesda, Maryland 20894, USA
{carmel, rogozin, wolf, koonin}@ncbi.nlm.nih.gov

Abstract. We propose a detailed model of evolution of exon-intron structure of eukaryotic genes that takes into account gene-specific intron gain and loss rates, branch-specific gain and loss coefficients, invariant sites incapable of intron gain, and rate variability of both gain and loss which is gamma-distributed across sites. We develop an expectation-maximization algorithm to estimate the parameters of this model, and study its performance using simulated data.

1 Introduction

Spliceosomal introns are one of the most prominent idiosyncrasies of eukaryotic genomes. They are scattered all over the eukaryota superkingdom, including, notably, species that are considered basal eukaryotes, such as *Giardia lamblia* [1]. This suggests that evolution of introns is intimately entangled with eukaryotic evolution; thus, the study of evolution of exon-intron structure of eukaryotic genes, apart from being interesting in its own right, might shed some light on the still enigmatic rise of eukaryotes. For example, one of the notorious, long-lasting unresolved issues in evolution of eukaryotic genomes is the intron-early versus intron-late debate. Proponents of the intron-early hypothesis posit that introns were prevalent at the earliest stages of cellular evolution and played a crucial role in the formation of complex genes via the mechanism of exon shuffling [2]. These introns were inherited by early eukaryotes but have been eliminated from prokaryotic genomes as a result of selective pressure for genome streamlining. By contrast, proponents of the intron-late hypothesis hold the view that introns had emerged, de novo, in early eukaryotes, and subsequent evolution of eukaryotes involved extensive insertion of new introns (see, e.g., [3,4]).

Various anecdotal studies have demonstrated certain features of intron evolution. But it was not until the accumulation of genomic information in the recent years that large-scale analyses became feasible. Such analyses yielded at least three different models of intron evolution. One model assumes parsimonious evolution [5]; another assumes a simple gene-specific gain/loss model and

A. McLysaght et al. (Eds.): RECOMB 2005 Ws on Comparative Genomics, LNBI 3678, pp. 35–46, 2005.
© Springer-Verlag Berlin Heidelberg 2005

analyzes it using Bayesian learning [6]; and yet another one assumes a simple branch-specific gain/loss model on three-species phylogenetic topology and analyzes it using direct maximum likelihood [7]. It seems that none of these models is sufficiently general, and each neglects different aspects of this complex evolutionary process. This is reflected in the major contradictions between the predictions laid out by the three models. For example, the gene-specific model [6] predicts an intron-poor eukaryotic ancestor and a dominating intron gain process; the branch-specific model [7] predicts an intron-rich eukaryotic ancestor and a dominating loss process; while the parsimonious model [5] is somewhat in between, predicting intermediate densities of introns in early eukaryotes, and a gain-dominated kaleidoscope of gain and loss events.

Here, we introduce a model of evolution of exon-intron structure, which is considerably more realistic than previously proposed models. The model accounts for gene-specific intron gain/loss mechanisms, branch-specific gain/loss mechanisms, invariant sites (a fraction of sites that are incapable of intron gain), and rate distribution across sites of both intron-gain and intron-loss. Using data from extant species, we follow the popular approach of estimating the model parameters by way of maximum likelihood. Direct maximization of the likelihood is, however, intractable in this case due to a large number of hidden random variables in the model. These are exactly the circumstances under which the expectation-maximization (EM) algorithm for maximizing the likelihood might prove itself useful. None of the software packages that we are aware of, either using direct maximization or EM, can deal with our proposed model. Hence, we devised an EM algorithm tailored to our particular model. As this model is rather detailed, a variety of biologically-reasonable models can be derived as special cases. For this reason, we anticipate a broad range of applicability to our algorithm, beyond its original use. In the following we describe our model of exon-intron structure evolution and an EM algorithm for learning its parameters.

2   The Evolutionary Model

Suppose that we have multiple alignments of G different genes from S eukaryotic species, and let our observed data be the projection, upon the above alignments, of a presence-absence intron map. That is, at every site in each species we can observe either zero (absence of an intron), one (presence of an intron), or $\star$ (missing value, indicating lack of knowledge about intron's presence or absence). Let us define a *pattern* as any column in an alignment, and let $\Omega \leq 3^S$ be the total number of unique observed patterns, indexed as $\omega_1, \ldots, \omega_\Omega$. We shall use n_{gp} to denote the number of patterns ω_p that are observed in gene g.

Let the rooted phylogeny of the above S species be given by an N-node binary tree, where $S = (N + 1)/2$. Let $q_0, \ldots, q_{N-1}$ be the nodes of this tree, with the convention that q_0 is the root node. We use the notations q^L, q^R and q^P to describe the left-descendant, right-descendant and parent, respectively, of node q (left and right are set arbitrarily). Also, let $\mathcal{L}(q)$ stand for the set of terminal nodes (leaves) that are descendants of q. We index the branches of the tree by the node into which they lead, and use Δ_q for the length of the branch (in time units) leading into node q. Hereinafter, we assume that the tree topology, as well as the branch lengths $\Delta_1, \ldots, \Delta_{N-1}$, are known.

Assume that the root node has a prior probability π_i of being at state i $(i = 0, 1)$, and that the transition matrix for gene g along branch t, $A_{ij}^g(q_t) = P(q_t = j | q_t^P = i)$, is described by

$$A^g(q_t) = \begin{pmatrix} 1 - \xi_t(1 - e^{-\eta_g \Delta_t}) & \xi_t(1 - e^{-\eta_g \Delta_t}) \\ 1 - (1 - \phi_t)e^{-\theta_g \Delta_t} & (1 - \phi_t)e^{-\theta_g \Delta_t} \end{pmatrix}, \tag{1}$$

where η_g and θ_g are gene-specific gain and loss rates, respectively, and ξ_t and ϕ_t are branch-specific gain and loss coefficients, respectively.

The common practice in evolutionary studies is to incorporate rate distribution across sites by associating each site with a *rate coefficient*, r, which scales the branch lengths of the corresponding phylogenetic tree, $\Delta_t \leftarrow r \cdot \Delta_t$. This rate coefficient is drawn from a probability distribution with non-negative domain and unit mean, typically the unit-mean gamma distribution. Such an approach is compatible with the notion that each site has a characteristic evolutionary rate. This, however, should be modified for intron evolution, where the gain and loss processes do not seem to be correlated. That is, sites that are fast to gain introns are not necessarily fast to lose them, and vice versa. Therefore, we model rate variation using two independent rate coefficients, r^η and r^θ, such that $\eta_g \leftarrow r^\eta \cdot \eta_g$ and $\theta_g \leftarrow r^\theta \cdot \theta_g$. These rates are independently drawn from the two distributions

$$r^\eta \sim \nu\delta(\eta) + (1 - \nu)\Gamma(\eta; \lambda) \tag{2}$$
$$r^\theta \sim \Gamma(\theta; \lambda).$$

Here, $\Gamma(x; \lambda)$ is the unit-mean gamma distribution of variable x with shape parameter λ, $\delta(x)$ is the Dirac delta-function, and ν is the fraction of sites that are invariant to gain (i.e., sites that are incapable of gaining introns). Two comments are in order with respect to these rate distributions. First, a site can be invariant only with respect to gain, in accord with the proto-splice site hypothesis that presumes preferential gain of introns at distinct sites [8]. In contrast, once an intron is gained, it can always be lost. Second, we assumed the same shape parameter for the gamma distributions of both gain and loss. This is done solely to simplify the already complex model. At a later stage, we may consider extending the model to include different shape parameters.

3   The EM Algorithm

Phylogenetic trees can be interpreted as Bayesian networks that depict an underlying evolutionary probabilistic model. Accordingly, the terminal nodes form the observed random variables of the model, and the internal nodes form the hidden random variables. Under this view, estimating the model parameters using EM is natural. Indeed, different EM algorithms have been applied to phylogenetic trees with various purposes [9–11]. The algorithm that resembles the one de-scribed here most closely was developed by Siepel & Haussler [12] and used for branch length optimization and parameter estimation of time-continuous Marko-vian processes. However, our model does not fit into any of the existing schemes as it includes several unique properties, such as the branch-specific coefficients, the gain-invariant sites, and the different treatment of rate variability across sites. In the rest of this section, we develop the algorithm in the context of the proposed model; we attempt to do so using notations that are as general as possible, in order to allow the use of this algorithm with other models as well.

Denote by $\mathcal{N}_g = (n_{1g}, \ldots, n_{\Omega g})$ the counts of all observed patterns in the gth alignment, and by Θ the set of model parameters. We will use, whenever necessary, the decomposition $\Theta = (\Xi, \Psi, \Lambda)$ where $\Xi = (\Xi_1, \ldots, \Xi_{N-1})$ is the set of branch-specific parameters, $\Xi_t = (\xi_t, \phi_t)$ in our case, characterized by not being affected by the rate variability; $\Psi = (\Psi_1, \ldots, \Psi_G)$ is the set of gene-specific variables, $\Psi_g = (\eta_g, \theta_g)$ in our case, characterized by being subject to rate variability, and $\Lambda = (\nu, \lambda)$ is the set of rate variables. We assume independence between genes and between sites, hence the likelihood function is

$$L(\mathcal{N}_1, \ldots, \mathcal{N}_G | \Theta) = \prod_{g=1}^{G} L(\mathcal{N}_g | \Xi, \Psi_g, \Lambda) = \prod_{g=1}^{G} \prod_{p=1}^{\Omega} L(\omega_p | \Xi, \Psi_g, \Lambda)^{n_{gp}}, \quad (3)$$

and the log-likelihood is just

$$\log L(\mathcal{N}_1, \ldots, \mathcal{N}_G | \Theta) = \sum_{g=1}^{G} \sum_{p=1}^{\Omega} n_{gp} \log L(\omega_p | \Xi, \Psi_g, \Lambda). \quad (4)$$

To make the rate distributions (2) amenable to *in silico* manipulations, we rendered them discrete as was done previously by Yang [13], using K cate-gories for the gamma distribution, and an additional category for the invariant sites. For the time being, we will keep our notations general and will not spec-ify the rendering technique, and in particular, will not assume equi-probable categories. Accordingly, r^θ can take the values $(r_1^\theta, \ldots, r_K^\theta)$ with probabilities $(f_1^\theta, \ldots, f_K^\theta)$, and r^η can take the values $(r_1^\eta = 0, r_2^\eta, \ldots, r_{K+1}^\eta)$ with probabili-ties $(f_1^\eta = \nu, f_2^\eta, \ldots, f_{K+1}^\eta)$. Introducing rate variability across sites is equivalent to transforming the model into a mixture model, with the rates determining the mixture coefficients. Consequently, we will associate with each site two discrete random variables, ρ_p^η and ρ_p^θ, indicating the rate category of η and θ, respec-tively. According to the EM paradigm, we are guaranteed to climb up-hill in $\log L(\omega_p | \Xi, \Psi_g, \Lambda)$, if we maximize the auxiliary function

$$Q_{gp}(\Xi, \Psi_g, \Lambda, \Xi^0, \Psi_g^0, \Lambda^0) = \quad (5)$$

$$= \sum_{\sigma, \rho_p^\eta, \rho_p^\theta} P(\sigma, \rho_p^\eta, \rho_p^\theta | \omega_p, \Xi^0, \Psi_g^0, \Lambda^0) \log P(\omega_p, \sigma, \rho_p^\eta, \rho_p^\theta | \Xi, \Psi_g, \Lambda) =$$

$$= \sum_{\sigma, \rho_p^\eta, \rho_p^\theta} P(\sigma, \rho_p^\eta, \rho_p^\theta | \omega_p, \Xi^0, \Psi_g^0, \Lambda^0) \cdot$$

$$\cdot \sum_{k=1}^{K+1} \sum_{k'=1}^{K} 1_{\{\rho_p^\eta = k\}} 1_{\{\rho_p^\theta = k'\}} \left[\log f_k^\eta + \log f_{k'}^\theta + \log P(\omega_p, \sigma | \Xi, \Psi_{gkk'}) \right].$$

Here, σ is any realization of the internal nodes of the tree, $1_{\{\rho = k\}}$ is a function that takes the value 1 when $\rho = k$ and takes the value zero otherwise, and $\Psi_{gkk'}$ is the set of effective gene-specific rates which, in our model, is $\Psi_{gkk'} = (\eta_{gk}, \theta_{gk'})$, where we have introduced the notations $\eta_{gk} = r_k^\eta \cdot \eta_g$ and $\theta_{gk'} = r_{k'}^\theta \cdot \theta_g$. If we now use

$$P(\sigma, \rho_p^\eta = k, \rho_p^\theta = k' | \omega_p, \Xi^0, \Psi_g^0, \Lambda^0) = \tag{6}$$
$$= P(\rho_p^\eta = k, \rho_p^\theta = k' | \omega_p, \Xi^0, \Psi_g^0, \Lambda^0) \cdot P(\sigma | \omega_p, \Xi^0, \Psi_{gkk'}^0)$$

in (5), we get

$$Q_{gp}(\Xi, \Psi_g, \Lambda, \Xi^0, \Psi_g^0, \Lambda^0) = \tag{7}$$

$$\sum_{k=1}^{K+1} \sum_{k'=1}^{K} \left[\sum_{\rho_p^\eta, \rho_p^\theta} P(\rho_p^\eta, \rho_p^\theta | \omega_p, \Xi^0, \Psi_g^0, \Lambda^0) \cdot 1_{\{\rho_p^\eta = k\}} 1_{\{\rho_p^\theta = k'\}} \right] \cdot$$

$$\cdot \left[\sum_{\sigma} P(\sigma | \omega_p, \Xi^0, \Psi_{gkk'}^0) \left[\log f_k^\eta + \log f_{k'}^\theta + \log P(\omega_p, \sigma | \Xi, \Psi_{gkk'}) \right] \right].$$

Denoting by $w_{gpkk'}$ and $Q_{gpkk'}$ the first and second square brackets, respectively, the auxiliary function maximization of which assures increasing the likelihood is

$$Q = \sum_{g=1}^{G} \sum_{p=1}^{\Omega} \sum_{k=1}^{K+1} \sum_{k'=1}^{K} n_{gp} w_{gpkk'} Q_{gpkk'}. \tag{8}$$

3.1   The E-Step

Here is how we compute $w_{gpkk'}$ and $Q_{gpkk'}$ for the current estimate Θ^0 of the model parameters.

$$w_{gpkk'} = \sum_{\rho_p^\eta, \rho_p^\theta} P(\rho_p^\eta, \rho_p^\theta | \omega_p, \Xi^0, \Psi_g^0, \Lambda^0) 1_{\{\rho_p^\eta = k\}} 1_{\{\rho_p^\theta = k'\}} = \tag{9}$$

$$= P(\rho_p^\eta = k, \rho_p^\theta = k' | \omega_p, \Xi^0, \Psi_g^0, \Lambda^0) =$$

$$= \frac{P(\rho_p^\eta = k | \Xi^0, \Psi_g^0, \Lambda^0) \cdot P(\rho_p^\theta = k' | \Xi^0, \Psi_g^0, \Lambda^0) \cdot P(\omega_p | \Xi^0, \Psi_{gkk'}^0)}{\sum_{h,h'} P(\rho_p^\eta = h | \Xi^0, \Psi_g^0, \Lambda^0) \cdot P(\rho_p^\theta = h' | \Xi^0, \Psi_g^0, \Lambda^0) \cdot P(\omega_p | \Xi^0, \Psi_{ghh'}^0)} =$$

$$= \frac{(f_k^\eta)^0 (f_{k'}^\theta)^0 P(\omega_p | \Xi^0, \Psi_{gkk'}^0)}{\sum_{h,h'} (f_h^\eta)^0 (f_{h'}^\theta)^0 P(\omega_p | \Xi^0, \Psi_{ghh'}^0)}.$$

The function $P(\omega_p|\Xi^0, \Psi^0_{gkk'})$ is the likelihood of the tree that we rapidly compute using a variant of Felsenstein's pruning algorithm [14]. To this end, let us define $\gamma^{gpkk'}(q) = P(\mathcal{L}(q)|q^P, \Xi^0, \Psi^0_{gkk'})$, which is the probability of observing those terminal nodes that are descendants of q, for a given state of the parent of q. Omitting the superscripts for clarity, this function is initialized at all terminal nodes $q_t \in \mathcal{L}(q_0)$ by

$$
\gamma(q_t) = \begin{cases} \begin{pmatrix} 1 - \xi_t(1 - e^{-\eta_{gk}\Delta_t}) \\ 1 - (1 - \phi_t)e^{-\theta_{gk'}\Delta_t} \end{pmatrix} & s_t = 0 \\[2mm] \begin{pmatrix} \xi_t(1 - e^{-\eta_{gk}\Delta_t}) \\ (1 - \phi_t)e^{-\theta_{gk'}\Delta_t} \end{pmatrix} & s_t = 1, \end{cases} \tag{10}
$$

where s_t is the value observed at q_t. Then, γ is computed at all internal nodes (except for the root) using the inward-recursion

$$
\gamma_i(q_t) = \sum_{j=0}^{1} A^g_{ij}(q_t)\tilde{\gamma}_j(q_t), \tag{11}
$$

where $\tilde{\gamma}_j(q)$ is an abbreviation for $\gamma_j(q^L)\gamma_j(q^R)$. The likelihood of the tree is then

$$
P(\omega_p|\Xi^0, \Psi^0_{gkk'}) = \sum_{i=0}^{1} \pi_i \tilde{\gamma}_i(q_0). \tag{12}
$$

Using this in (9) allows us to compute the coefficients $w_{gpkk'}$. In order to compute the coefficients $Q_{gpkk'}$ we need a complementary recursion to the above γ-recursion. To this end, let us define $\alpha^{gpkk'}(q, q^P) = P(q, q^P|\omega_p, \Xi^0, \Psi^0_{gkk'})$. Again, omitting the superscripts, this function can be initialized on the two descendants of the root by

$$
\alpha(q, q_0) = \frac{1}{P(\omega_p|\Xi^0, \Psi^0_{gkk'})} \begin{cases} \begin{pmatrix} \pi_0\gamma_0(q^S)A^g_{00}(q)\, 0 \\ \pi_1\gamma_1(q^S)A^g_{10}(q)\, 0 \end{pmatrix} & q \in \mathcal{L}(q_0), \quad s = 0 \\[2mm] \begin{pmatrix} 0\; \pi_0\gamma_0(q^S)A^g_{01}(q) \\ 0\; \pi_1\gamma_1(q^S)A^g_{11}(q) \end{pmatrix} & q \in \mathcal{L}(q_0), \quad s = 1 \\[2mm] \begin{pmatrix} \pi_0\gamma_0(q^S)\tilde{\gamma}_0(q)A^g_{00}(q)\; \pi_0\gamma_0(q^S)\tilde{\gamma}_1(q)A^g_{01}(q) \\ \pi_1\gamma_1(q^S)\tilde{\gamma}_0(q)A^g_{10}(q)\; \pi_1\gamma_1(q^S)\tilde{\gamma}_1(q)A^g_{11}(q) \end{pmatrix} & q \notin \mathcal{L}(q_0). \end{cases} \tag{13}
$$

Here, q is a descendent of the root (either q_0^R or q_0^L), and q^S is its sibling. For any other internal node, α is computed using the outward-recursion

$$
\alpha(q, q^P) = \begin{pmatrix} \frac{\tilde{\gamma}_0(q)}{\gamma_0(q)}\beta_0(q^P)A^g_{00}(q)\; \frac{\tilde{\gamma}_1(q)}{\gamma_0(q)}\beta_0(q^P)A^g_{01}(q) \\ \frac{\tilde{\gamma}_0(q)}{\gamma_1(q)}\beta_1(q^P)A^g_{10}(q)\; \frac{\tilde{\gamma}_1(q)}{\gamma_1(q)}\beta_1(q^P)A^g_{11}(q) \end{pmatrix}, \tag{14}
$$

where $\beta(q) = P(q|\omega_p, \Xi^0, \Psi^0_{gkk'}) = \sum_{q^P} \alpha(q, q^P)$ is computed for each node subsequently to the computation of α. Finally, for each terminal node that is not a descendant of the root,

$$\alpha(q, q^P) = \begin{cases} \begin{pmatrix} \beta_0(q^P) \ 0 \\ \beta_1(q^P) \ 0 \\ 0 \ \beta_0(q^P) \\ 0 \ \beta_1(q^P) \end{pmatrix} & s = 0 \\ & \\ \begin{pmatrix} 0 \ \beta_0(q^P) \\ 0 \ \beta_1(q^P) \end{pmatrix} & s = 1. \end{cases} \tag{15}$$

This inward-outward recursion is the phylogenetic equivalent of the backward-forward recursion known from hidden Markov models, and other versions of it have already been developed, see, e.g., [9,12]. We shall now see how the α's and β's allow us to compute the coefficients $Q_{gpkk'}$. Notice that, if we use the state variables as indices, we can replace the function $\log P(\omega_p, \sigma | \Xi, \Psi_{gkk'})$ in (7) by

$$\log P(\omega_p, \sigma | \Xi, \Psi_{gkk'}) = \sum_{i=0}^{1} (q_0)_i \log \pi_i + \sum_{i,j=0}^{1} \sum_{t=1}^{N-1} (q_t)_j (q_t^P)_i \log A_{ij}^g(q_t). \tag{16}$$

Denote the expectation over $P(\sigma | \omega_p, \Xi^0, \Psi_{gkk'}^0)$ by E_σ. Applying it to (16) we get

$$E_\sigma \left[\log P(\omega_p, \sigma | \Xi, \Psi_{gkk'}) \right] = \tag{17}$$

$$= \sum_{i=0}^{1} \log \pi_i E_\sigma [(q_0)_i] + \sum_{i,j=0}^{1} \sum_{t=1}^{N-1} \log A_{ij}^g(q_t) E_\sigma [(q_t)_j (q_t^P)_i].$$

But, $E_\sigma[(q_0)_i] = P(q_0 = i | \omega_p, \Xi^0, \Psi_{gkk'}^0) = \beta_i(q_0)$, and similarly $E_\sigma[(q_t)_j(q_t^P)_i] = \alpha_{ij}(q_t, q_t^P)$, so that $Q_{gpkk'}$ can be finally written as

$$Q_{gpkk'} = \sum_\sigma P(\sigma | \omega_p, \Xi^0, \Psi_{gkk'}^0) \cdot \tag{18}$$

$$\cdot \left[\log f_k^\eta + \log f_{k'}^\theta + \log P(\omega_p, \sigma | \Xi, \Psi_{gkk'}) \right] =$$

$$= \log f_k^\eta + \log f_{k'}^\theta + \sum_{i=0}^{1} \beta_i(q_0) \log \pi_i + \sum_{i,j=0}^{1} \sum_{t=1}^{N-1} \alpha_{ij}(q_t, q_t^P) \log A_{ij}^g(q_t).$$

One of the appealing features of EM is that is allows, in many cases, to treat missing data fairly easily. In our case, two simple modifications are required for this. Firstly, we have to add to the γ-recursion initialization (10) an option

$$\gamma(q_t) = \begin{pmatrix} 1 \\ 1 \end{pmatrix} \qquad s_t = \star. \tag{19}$$

Secondly, we have to add to the α-recursion finalization (15) an option

$$\alpha(q_t) = \begin{pmatrix} \beta_0(q_t^P) A_{00}^g(q_t) \ \beta_0(q_t^P) A_{01}^g(q_t) \\ \beta_1(q_t^P) A_{10}^g(q_t) \ \beta_1(q_t^P) A_{11}^g(q_t) \end{pmatrix} \qquad s_t = \star. \tag{20}$$

3.2   The M-Step

Substituting the expressions for $w_{gpkk'}$ and $Q_{gpkk'}$ in (8), we obtain the final form of the function to be maximized at each iteration. Explicitly, this is

$$Q = \sum_{g=1}^{G}\sum_{p=1}^{\Omega}\sum_{k=1}^{K+1}\sum_{k'=1}^{K} n_{gp}w_{gpkk'}(\log f_k^{\eta} + \log f_{k'}^{\theta}) + \tag{21}$$

$$+ \sum_{g=1}^{G}\sum_{p=1}^{\Omega}\sum_{k=1}^{K+1}\sum_{k'=1}^{K} n_{gp}w_{gpkk'}\left[\beta_0^{gpkk'}(q_0)\log\pi_0 + \beta_1^{gpkk'}(q_0)\log\pi_1\right] +$$

$$+ \sum_{g=1}^{G}\sum_{p=1}^{\Omega}\sum_{k=1}^{K+1}\sum_{k'=1}^{K}\sum_{t=1}^{N-1} n_{gp}w_{gpkk'}\alpha_{00}^{gpkk'}(q_t)\log\left[1 - \xi_t(1 - e^{-\eta_{gk}\Delta_t})\right] +$$

$$+ \sum_{g=1}^{G}\sum_{p=1}^{\Omega}\sum_{k=1}^{K+1}\sum_{k'=1}^{K}\sum_{t=1}^{N-1} n_{gp}w_{gpkk'}\alpha_{01}^{gpkk'}(q_t)\left[\log\xi_t + \log(1 - e^{-\eta_{gk}\Delta_t})\right] +$$

$$+ \sum_{g=1}^{G}\sum_{p=1}^{\Omega}\sum_{k=1}^{K+1}\sum_{k'=1}^{K}\sum_{t=1}^{N-1} n_{gp}w_{gpkk'}\alpha_{10}^{gpkk'}(q_t)\log\left[1 - (1 - \phi_t)e^{-\theta_{gk'}\Delta_t}\right] +$$

$$+ \sum_{g=1}^{G}\sum_{p=1}^{\Omega}\sum_{k=1}^{K+1}\sum_{k'=1}^{K}\sum_{t=1}^{N-1} n_{gp}w_{gpkk'}\alpha_{11}^{gpkk'}(q_t)\left[\log(1 - \phi_t) - \theta_{gk'}\Delta_t\right].$$

It is well-known that any increase in Q suffices to climb up-hill in the likelihood, and therefore it is not of utmost importance to maximize it precisely. Hence, we do not invest too much in finding precise maximum, but rather use low-tolerance maximization with respect to each of the parameters individually. Since it is easy to differentiate Q twice with respect to all the parameters (except for λ), we use the Newton-Raphson zero-finding algorithm for the maximization. Due to space limitations and because the derivation is, essentially, trivial, we do not present them here.

We must, however, devote a few words to the maximization of Q with respect to λ. In (21) we kept the rate distributions general, but (2) imposes the constraints $r_k^{\theta} = r_{k+1}^{\eta}$. Furthermore, in rendering the gamma distribution discrete, we assume equi-probable categories, thus

$$f_{k+1}^{\eta} = (\nu - 1)f_k^{\theta} = \frac{\nu - 1}{K} \qquad k = 1, \ldots, K. \tag{22}$$

Therefore, Q depends on λ through r^{η} and r^{θ}, making analytic differentiation impossible. Thus, in this case, we used Brent's maximization algorithm that does not require derivatives.

4   Validation

We intend to apply the algorithm to real data, namely, an amended version of the data set from [5], which consists of multiple alignments of over 700 orthologous genes from 8 eukaryotic species. However, prior to its application to real data, the algorithm must be carefully validated against simulated data. Thus, we have written simulation software that performs three tasks. Firstly, given the number of extant species, it builds a random phylogeny. Secondly, it assigns random lengths to the branches based on the exponential distribution (keeping the tree balanced). Thirdly, it draws the model parameters subject to some biologically plausible constraints. Given the phylogenetic tree and the model parameters, we then simulate any desired number of evolutionary scenarios, collecting the observations on the terminal nodes.

While EM algorithms always converge to a maximum of the likelihood, they are not guaranteed to find the global maximum. In practice, however, we have strong indications that our EM algorithm is highly effective in finding the global maximum. We cannot provide a proof for this, but at least it is clear that it always estimates model parameters that give a higher likelihood than the true model parameters, see Figure 1.

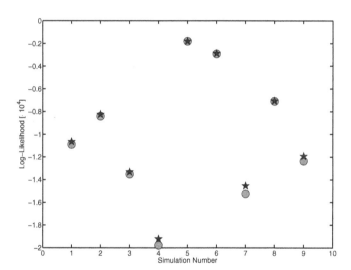

Fig. 1. Summary of 9 independent simulations. For each simulation, a 4 species random phylogeny spanning 400 million years and a set of model parameters were drawn randomly. Intron evolution was simulated for four multigenes of mean length of 5000 AA, with no rate variation. Parameters were estimated using tolerance of 10^{-2}. The dots indicate log-likelihood values computed for the true model parameters, and the pentagons indicate log-likelihood values computed for the estimated parameters. Note that the log-likelihood of the estimated parameters is always greater than that of the true parameters.

A well known property of maximum likelihood estimators is that they are not guaranteed to be unbiased for any finite sample size. In our model, and

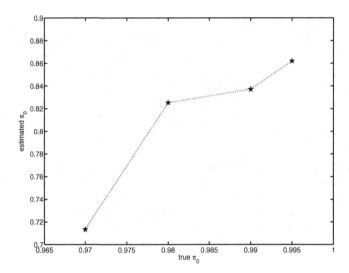

Fig. 2. Estimated π_0 versus true π_0. Each dot is the mean of three simulations of six-species random phylogeny spanning 300 million years. In each simulation, we assumed four multigenes of mean length of 50,000 AA, with no rate variation.

probably in other phylogenetic models, the bias might be significant, mainly due to the small number of species and to the paucity of informative patterns. An example is shown in Figure 2, where the probability π_0 of the root node is estimated. This problem can be less severe when a monotonic relation between the true parameter and the estimated one holds (Figure 2). We are currently investigating different approaches to map this bias more accurately.

5   Discussion

We describe here an algorithm that allows for parameter estimation of an evolutionary model for exon-intron structure of eukaryotic genes. Once estimated, these parameters could help resolving the current debate regarding evolution of introns, in particular, with regard to the relative contributions of intron loss and gain in different eukaryotic lineages.

Some of the assumptions of our model are worth discussion. Specifically, in Equations (3) and (4), we assumed that different sites evolve (i.e., gain and lose introns) independently. However, several observations show that such independence is only an approximation. First, introns in intron-poor species tend to cluster near the 5' end of the gene [15,16]. Second, adjacent introns tend to be lost in concert [16,17]. Nevertheless, it seems that such site-dependence of gain and loss is a secondary factor in intron evolution. First, non-homogeneous spatial distribution of introns along the gene is pronounced only in species with a low number of introns. Second, some anecdotal studies could not find any preference of adjacent introns to be lost together (e.g., [18].) Should subsequent studies indicate that the dependence between sites is more important than we currently envisage, our model probably can be extended using the context-dependent ideas developed in [12].

Similarly, in Equations (3) and (4), we assumed that different genes gain and lose introns independently. Currently, we are unaware of any strong evidence for such dependence, but if it is discovered, it can be easily accounted for in our model by concatenating genes with similar evolutionary trends and treating them as a single multigene.

Additionally, we assumed the same shape parameter for the gamma distribution of intron gain and loss rates. As mentioned above, this assumption was taken out of convenience, and due to the general impression that the exact shape of the gamma distribution is not a primary factor. However, our model can be rather easily extended to incorporate different shape parameters for gain and loss.

The computational complexity of the algorithm is, in the worst case, $O(G \cdot S \cdot K^2 \cdot 3^S)$. The exponential dependency arises because the number of unique patterns, Ω, is exponential with the number of species. However, if W is the total number of sites in all the alignments, it bounds Ω by $\Omega \leq \min(W, 3^S)$, thus keeping us, in practice, far away off the worst case.

The current Matlab® code is too slow to handle efficiently the real data (over two million sites) and the massive simulations. Therefore, we are in the process of writing the code in C++, allowing for its application to large data sets. The C++ software will be made available as soon as it is ready.

Acknowledgements

The authors would like to thank Jacob Goldberger for useful discussions regarding aspects of the EM algorithm.

References

1. Nixon, J. E., Wang, A., Morrison, H.G., McArthur, A.G., Sogin, M.L., Loftus, B.J., Samuelson, J.: A Spliceosomal Intron in Giardia Lamblia. Proc. Natl. Acad. Sci. USA **99** (2002) 3701–3705.
2. Gilbert, W.: The Exon Theory of Genes. Cold Spring Harb. Symp. Quant. Biol. **52** (1987) 901–905.
3. Cho, G., Doolittle, R.F.: Intron Distribution in Ancient Paralogs Supports Random Insertions and Not Random Loss. J. Mol. Evol. **44** (1997) 573–584.
4. Lynch, M.: Intron Evolution as a Population-genetic Process. Proc. Natl. Acad. Sci. USA **99** (2002) 6118–6123.
5. Rogozin, I.B., Wolf, Y.I., Sorokin, A.V., Mirkin, B.G., Koonin, E.V.: Remarkable Interkingdom Conservation of Intron Positions and Massive, Lineage-Specific Intron Loss and Gain in Eukaryotic Evolution. Curr. Biol. **13** (2003) 1512–1517.
6. Qui, W.-G., Schisler, N., Stoltzfus, A.: The Evolutionary Gain of Spliceosomal Introns: Sequence and Phase Preferences. Mol. Biol. Evol. **21** (2004) 1252–1263.

7. Roy, S.W., Gilbert, W.: Complex Early Genes. Proc. Natl. Acad. Sci. USA **102** (2005) 1986–1991.
8. Dibb, N.J.: Proto-Splice Site Model of Intron Origin. J. Theor. Biol. **151** (1991) 405–416.
9. Friedman, N., Ninio, M., Pe'er I., Pupko, T.: A Structural EM Algorithm for Phylogenetic Inference. J. Comput. Biol. **9** (2002) 331–353.
10. Holmes, I.: Using Evolutionary Expectation Maximisation to Estimate Indel Rates. Bioinformatics **21** (2005) 2294–2300.
11. Brooks, D. J., Fresco, J. R., Singh, M.: A Novel Method for Estimating Ancestral Amino Acid Composition and Its Application to Proteins of the Last Universal Ancestor. Bioinformatics **20** (2004) 2251–2257.
12. Siepel, A., Haussler, D.: Phylogenetic Estimation of Context-Dependent Substitution Rates by Maximum Likelihood. Mol. Biol. Evol. **21** (2004) 468–488.
13. Yang, Z.: Maximum Likelihood Phylogenetic Estimation from DNA Sequences with Variable Rates over Sites: Approximate Methods. J. Mol. Evol. **39** (1994) 306–314.
14. Felsenstein, J.: Evolutionary Trees from DNA Sequences: A Maximum Likelihood Approach. J. Mol. Evol. **17** (1981) 368–376.
15. Mourier, T, Jeffares, D.C.: Eukaryotic Intron Loss. Science **300** (2003) 1393.
16. Sverdlov, A.V., Babenko, V.N., Rogozin, I.B., Koonin, E.V.: Preferential Loss and Gain of Introns in 3' Portions of Genes Suggests a Reverse-Transcription Mechanism of Intron Insertion. Gene **338** (2004) 85–91.
17. Roy, S.W., Gilbert, W.: The Pattern of Intron Loss. Proc. Natl. Acad. Sci. USA **102** (2005) 713–718.
18. Cho, S., Jin, S.-W., Cohen, A., Ellis, R.E.: A Phylogeny of Caenorhabditis Reveals Frequent Loss of Introns During Nematode Evolution. Genome Res. **14** (2004) 1207–1220.

Likely Scenarios of Intron Evolution

Miklós Csűrös

Department of Computer Science and Operations Research,
Université de Montréal,
C.P. 6128, succ. Centre-Ville, Montréal, Qué., Canada, H3C 3J7
csuros@iro.umontreal.ca

Abstract. Whether common ancestors of eukaryotes and prokaryotes had introns is one of the oldest unanswered questions in molecular evolution. Recently completed genome sequences have been used for comprehensive analyses of exon-intron organization in orthologous genes of diverse organisms, leading to more refined work on intron evolution. Large sets of intron presence-absence data require rigorous theoretical frameworks in which different hypotheses can be compared and validated. We describe a probabilistic model for intron gains and losses along an evolutionary tree. The model parameters are estimated using maximum likelihood. We propose a method for estimating the number of introns lost or unobserved in all extant organisms in a study, and show how to calculate counts of intron gains and losses along the branches by using posterior probabilities. The methods are used to analyze the most comprehensive intron data set available presently, consisting of 7236 intron sites from eight eukaryotic organisms. The analysis shows a dynamic history with frequent intron losses and gains, and fairly — albeit not as greatly as previously postulated — intron-rich ancestral organisms.

1 Introduction

A major difference between eukaryotic and prokaryotic gene organization is that many eukaryotic genes have a mosaic structure: coding sequences are separated by intervening non-coding sequences, known as introns. Francis Crick's 1979 comment [1] on the evolutionary origins of spliceosomal introns — "I have noticed that this question has an extraordinary fascination for almost everybody concerned with the problem" — could have been said yesterday. The problem is still not completely resolved [2]. The question of whether or not the most recent common ancestor of eukaryotes and prokaryotes had introns, known as the "introns early/late" debate [3], is one of the oldest unanswered questions in molecular evolution. Recent advances [4–8] rely on whole-genome sequences for diverse organisms. It has become clear that introns have been gained and lost in different lineages at various rates. In this context it is of particular interest to estimate the intron densities in early eukaryotic organisms, as well as rates and patterns of intron loss and gain along different evolutionary lineages. The aim of this article is to describe a probabilistic model which allows for a maximum likelihood (ML) analysis of rates and scenarios. We describe some methods to this end and apply them to a data set of 7236 introns from eight fully sequenced eukaryotic organisms.

A. McLysaght et al. (Eds.): RECOMB 2005 Ws on Comparative Genomics, LNBI 3678, pp. 47–60, 2005.
© Springer-Verlag Berlin Heidelberg 2005

2   A Probabilistic Model for Intron Evolution

In order to model the evolution of introns along an evolutionary tree, we use a Markov model that permits varying rates along different branches, described as follows. Let T be a phylogenetic tree T over a set of species X: T is a rooted tree in which the leaves are bijectively labeled by the elements of X. Let $E(T)$ denote the set of edges (directed away from the root), and let $V(T)$ denote the node set of the tree. Throughout the paper, intron presence is encoded by the value 1, and intron absence is encoded by the value 0. Along each edge $e \in E(T)$, introns are generated by a two-state continuous-time Markov process with *gain* and *loss rates* $\lambda_e, \mu_e \geq 0$, respectively. The length of an edge e is denoted by t_e. In addition, the root is associated with the *root probabilities* π_0, π_1 with $\pi_0 + \pi_1 = 1$. The tree T with its parameters defines a stochastic evolution model for the *state* $\tilde{\chi}(u)$ of an intron site at every tree node $u \in V(T)$ in the following manner. The intron is present at the root with probability π_1. The intron state evolves along the tree edges from the root towards the leaves, and changes on each edge according to the transition probabilities. For every child node v and its parent u,
$$\mathbb{P}\Big\{\tilde{\chi}(v) = j \;\Big|\; \tilde{\chi}(u) = i\Big\} = p_{i \to j}(uv),$$
where $p_{i \to j}$ are determined by the edge parameters, which we discuss shortly. The values at the leaves form the *character* $\chi = (\tilde{\chi}(u) \colon u \in X)$. The input data set (or *sample*) consists of independent and identically distributed (iid) characters: $D = (\chi_i \colon i = 1, \dots, n)$.

Using standard results [9], the transition probabilities along the edge e with rates $\lambda_e = \lambda, \mu_e = \mu$ and length t can be written as

$$p_{0 \to 0}(e) = \frac{\mu}{\lambda + \mu} + \frac{\lambda}{\lambda + \mu} e^{-t(\lambda + \mu)} \qquad p_{0 \to 1}(e) = \frac{\lambda}{\lambda + \mu} - \frac{\lambda}{\lambda + \mu} e^{-t(\lambda + \mu)}$$

$$p_{1 \to 0}(e) = \frac{\mu}{\lambda + \mu} - \frac{\mu}{\lambda + \mu} e^{-t(\lambda + \mu)} \qquad p_{1 \to 1}(e) = \frac{\lambda}{\lambda + \mu} + \frac{\mu}{\lambda + \mu} e^{-t(\lambda + \mu)}.$$

In the absence of independent edge length estimates, we fix the scaling for the edge lengths in such a way that $\lambda_e + \mu_e = 1$.

A somewhat more complicated model of intron evolution was used by Rzhetsky *et al.* [10], who also accounted for possible intron sliding [11], whereby orthologous intron sites may differ by a few positions with respect to the underlying coding sequence in different organisms. In our case, the orthology criterion incorporates intron sliding a priori. Some other authors (e.g., [5]) imposed a reversible Markov model with identical rates across different branches, which is not entirely realistic for intron evolution, but nevertheless can result in important insights already.

3   ML Estimation of Parameters and Scenarios

3.1   Unobserved Intron Sites

Our goal is to design a maximum likelihood approach to estimate the model parameters on a given tree T, and to calculate likely scenarios of intron gains and losses along the edges. The described probabilistic model is fairly simple, and the parameters can be estimated from a data set by usual optimization techniques [12]. There is, however, an inherent difficulty in analyzing an intron absence/presence data set: there is no obvious evidence of introns lost in all extant organisms in the study. Consequently, one has access only to a sample of iid characters from which the all-0 characters ("unobserved introns") have been removed. Maximizing the likelihood without the all-0 characters introduces a bias. At the same time, it is not possible to estimate the number of missing all-0 characters by maximizing either the likelihood (every added all-0 character decreases it), or the average likelihood (an unbounded number of all-0 characters can be added if their likelihood is large enough). It is therefore necessary to separate the estimation of unobserved sites from likelihood maximization.

The problem of augmenting the data set with a certain number of all-0 characters has a particular relevance for the complexity of ML estimation of phylogenies. Tuffley and Steel [13] showed that ML and maximum parsimony (MP) yield the same optimal tree topology when enough all-0 characters are added to the data set in a symmetric binary model. Their result was employed very recently [14,15] to demonstrate the NP-hardness of ML optimization for phylogenies. The theoretical connection between ML and MP established by the addition of all-0 characters has direct practical consequences in the case of intron data sets. For instance, the analyses of the same sample carried out by two groups of researchers [4,16–18], using ML and MP, arrived at different conclusions concerning intron gain/loss rates and ancient intron density. Some of the disagreements can be attributed to different assumptions about unobserved sites, instead of methodological issues.

For a formal discussion, define the following notions. An *extension* $\tilde{\chi}$ of a character χ is an assignment of states to every tree node that agrees with χ at the leaves. Let $H(\chi)$ denote the set of all extensions of χ. The likelihood of a character χ is the probability

$$f_\chi = \sum_{\tilde{\chi} \in H(\chi)} \pi_{\tilde{\chi}(\text{root})} \prod_{uv \in E(T)} p_{\tilde{\chi}(u) \to \tilde{\chi}(v)}(uv).$$

The likelihood of a complete data set $D = (\chi_i : i = 1, \ldots, n)$ is simply $L(D) = \prod_{i=1}^{n} f_{\chi_i}$. Let f_0 denote the likelihood of the all-0 character $0^{|X|}$. The expected number of all-0 characters in a data set of size n is nf_0. Accordingly, the expected number of unobserved sites given that there are $\bar{n}$ observed ones (non-all-0 characters in the data set), is

$$\hat{n}_0 = \bar{n} \frac{f_0}{1 - f_0}. \tag{1}$$

(The distribution of the number of unobserved sites is a negative binomial distribution with parameters $\bar{n}$ and $(1 - f_0)$.)

Let $\bar{D} = (\chi_i : i = 1, \ldots, \bar{n})$ denote the observed sample, without the all-0 characters, and $n_0 = n - \bar{n}$ denote the true number of unobserved sites. Figure 1 sketches the algorithm GUESS-THE-SAMPLE for ML estimation of model parameters using a guess for n_0. The guess is used to optimize the model parameters and then to compute the expected number of unobserved sites using Eq. (1). Line G4 compares the latter with the original guess and if they differ too much, it rejects the optimized parameters. The exact definition of "too much" can rely on the concentration properties of n_0: for a given sample size n, it is binomially distributed with parameters n and f_0 with a variance of $nf_0(1 - f_0)$. For example, the guess Z can be rejected if

$$|\hat{n}_0 - Z| > c\sqrt{(\bar{n} + Z)f_0(1 - f_0)},$$

where c is a constant determining the desired confidence level. Figure 3 shows the behavior of this difference for a data set analyzed in Section 4. Notice that the plot suggests that n_0 could be estimated by an iterative technique, in which two steps are alternating: (1) estimation of the number of intron sites, based on model parameters, and observed introns, and (2) maximization of the likelihood given the estimated number of intron sites. In other words, $\hat{n}_0$ can be fed back to the algorithm in Line G4 in lieu of rejection, until convergence is reached. Based on the plot of Fig. 3, however, the convergence is very slow, and there is nothing gained over trying basically all possible values for n_0. (There is an upper bound given by the length of sequences from which $\bar{D}$ was obtained.)

Algorithm GUESS-THE-SAMPLE
Input A guess Z for n_0, observed sample $\bar{D} = (\chi_i : i = 1, \ldots, \bar{n})$
G1 Set $D' = (\chi_i' : i = 1, \ldots, \bar{n} + Z)$ with $\chi_i' = \chi_i$ for $i \le \bar{n}$ and $\chi_i' = 0^{|X|}$ for $i > \bar{n}$.
G2 Optimize the model parameters on the augmented sample D'.
G3 Calculate $\hat{n}_0$ by using the optimized model parameters in Eq. (1).
G4 Reject if $\hat{n}_0$ differs from Z by too much.

Fig. 1. ML parameter estimation with unknown number n_0 of unobserved sites

---

3.2   Patterns of Intron Gain and Loss Along Tree Edges

Once the number of unobserved intron sites is estimated and the model parameters are optimized, the model can be used to infer likely scenarios of intron evolution. In particular, exact posterior probabilities for intron presence can be calculated at each node, or for intron loss and gain on each branch. Define the *lower conditional likelihood* for every node u, site i, and state $x \in \{0, 1\}$ by:

$$L_i^{(x)}(u) = I\{x = \chi_i(u)\} \qquad \text{when } u \text{ is a leaf,}$$

$$L_i^{(x)}(u) = \prod_{v \in \text{children}(u)} \left(\sum_{y \in \{0,1\}} p_{x \to y}(uv) L_i^{(y)}(v) \right) \qquad \text{when } u \text{ is not a leaf,}$$

where $I\{A\}$ is the indicator function: $I\{A\} = 1$ if A is true, otherwise $I\{A\} = 0$. The value $L_i^{(x)}(u)$ is the probability of observing the states from character χ_i at the leaves of the subtree T_u rooted at u, given that u is in state x.

We also need the *upper conditional likelihood* $U_i^{(x)}(u)$, which is the probability of observing the states from character χ_i at leaves that are not in the subtree T_u, given that u is in state x. The upper conditional likelihoods can be computed by dynamic programming, using the following recursions in a breadth-first traversal.

$$U_i^{(x)}(\text{root}) = 1$$

$$U_i^{(x)}(u) = \sum_{y \in \{0,1\}} p_{y \to x}(vu) U_i^{(y)}(v) \prod_{w \in \text{siblings}(v)} \left(\sum_{z \in \{0,1\}} p_{y \to z}(vw) L_i^{(z)}(w) \right),$$

where v is the parent of u.

The posterior probability that node u is in state x at site i equals

$$q_i^{(x)}(u) \propto U_i^{(x)}(u) L_i^{(x)}(u).$$

Usual posterior calculations of ancestral states described in, e.g., [19,12] apply to reversible mutation models, when the tree can be rerooted at u and then $L^{(x)}(u)$ can be used directly. Here we need the additional technicality of computing upper conditional likelihoods. One can also compute the posterior probability of site i undergoing a $x \to y$ transition on the edge leading to the node v from its parent u as

$$q_i^{(x \to y)}(uv) \propto U_i^{(x)}(u) p_{x \to y}(uv) L_i^{(y)}(v).$$

Working with posterior probabilities instead of the single most likely extension has the advantage that posterior probabilities can be summed to obtain expected counts for intron gains and losses. The *posterior mean counts* of states at a node u, or state transitions $(x \to y)$ on an edge uv are computed as

$$n^{(x)}(u) = \sum_{i=1}^{n} q_i^{(x)}(u),$$

$$n^{(x \to y)}(uv) = \sum_{i=1}^{n} q_i^{(x \to y)}(uv), \tag{2}$$

respectively. (Notice that the sums include the unobserved intron sites.) In particular, $n^{(1)}(u)$ is the expected number of introns present at node u, given the model parameters and the observed data. Similarly, $n^{(0 \to 1)}(uv)$ is the expected number of introns gained, and $n^{(1 \to 0)}(uv)$ is the expected number of introns lost along the edge uv.

4   Intron Evolution in Eukaryotes

Rogozin *et al.* [4] compiled a data set based on orthologous protein groups in eukaryotic organisms. They aligned protein sequences with the genome sequences of eight fully sequenced organisms, and defined orthologous intron positions based on conserved regions in the alignments. The data set (downloaded from `ftp://ftp.ncbi.nlm.nih.gov/pub/koonin/intron_evolution`) consists of 7236 orthologous intron positions, from 684 protein groups. Figure 2 shows the organisms involved in the study, as well as the number of introns for each organism.

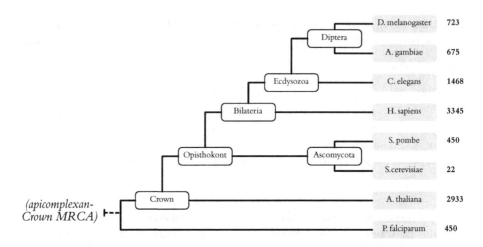

Fig. 2. Phylogenetic tree for the data set in Section 4, showing taxon names and intron counts. *P. falciparum* serves as an outgroup. Only the solid edges were used in the computations. The edge that connects *P. falciparum* to the tree accounts for changes between the Opisthokont node, and the most recent common ancestor (MRCA) of plants, animals, fungi, and apicomplexans, as well as for those leading from that MRCA to *P. falciparum*.

We note in passing that there is some ongoing debate [20–23] as to whether the phylogenetic tree of Fig. 2 is correct, namely, whether Ecdysozoa are monophyletic. Philippe *et al.* [22] argue that they are, and that support for other hypotheses are due to long branch attraction phenomena. Roy and Gilbert [21] also argue for an ecdysozoan clade, based on the intron data set of [4]. We consider only one phylogenetic tree, and leave further analysis to a more complete version of this abstract.

We implemented a Java package for the analysis of intron data sets, which performs parameter optimization and posterior calculations. As we indicated in §3.1, it is necessary to estimate the number of unobserved intron sites before proceeding to likelihood maximization. Figure 3 shows the estimation procedure applied to the data at hand. The estimation reaches a fix point at around 35 thousand unobserved characters, i.e., likelihood optimization with that many all-0 characters gives an equal expectation (within integer rounding) for the number of unobserved characters. Allowing for some statistical error, about 20–80 thousand unobserved characters give an expectation that is within twice the standard error after parameter optimization.

Using 35000 unobserved characters, we proceeded to parameter optimization, and then to the estimation of intron loss and gain patterns. Rogozin *et al.* [4] computed losses and gains using Dollo parsimony [24,25], assuming that every intron arose only once along the tree.

Roy and Gilbert [16,17] estimated transition probabilities and intron counts using "local" optimization, independently for each edge. (A similar method was

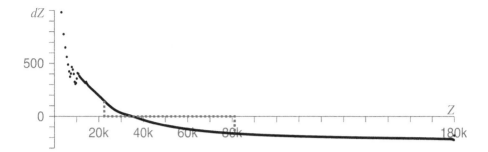

Fig. 3. Estimation of unobserved intron sites. The X axis shows the guess Z with which algorithm GUESS-THE-SAMPLE is invoked, and the Y axis shows the difference $\hat{n}_0 - Z$ calculated after parameter optimization. The dotted lines delineate the region in which the difference is below twice the standard deviation.

used in [6].) Their principal technique is a tree contraction, in which a whole sub-tree is replaced by a single branch, and the corresponding characters are derived by computing a logical OR over the intron states at the subtree leaves. They provide separate sets of formulas for analyzing exterior and interior branches. In the case of exterior branches, three-leaf star trees are formed, in which the original edge is preserved, a second edge is contracted from the sibling subtree, and the third edge is contracted from the rest of the tree. In the case of in-ternal branches, they contract the subtrees for the four neighbors of the edge endpoints to form a quartet. (The method applies only to binary trees.) The methods of [16,17] estimate a larger number of parameters than our likelihood optimization: in addition to the probabilities of intron inheritance, various intron loss and gain counts are independently estimated on each branch. It is plausible that by not enforcing consistency between different estimates that depend on the same parameter (for instance, the same edge transition probabilities should ap-pear in many different contractions), the results may get distorted. In addition, the Roy-Gilbert formulas do not account for the possibility of introns arising more than once.

Multiple origins of introns in an orthologous position are explicitly forbid-den by Dollo parsimony. Parallel gains are allowed in our probabilistic model, and may in truth account for a number of shared introns between eukaryotic kingdoms [5,18]. Even if one disregards for a moment the question of parallel gains, Dollo parsimony still has its own shortcomings when used for reconstruct-ing plausible histories. If intron gains are much less probable than intron losses, Dollo parsimony retrieves the most likely extension for every single character. It is not suitable, however, for determining cumulative values such as ancestral intron counts, since then the contribution of second, third, etc. most probable histories cannot be neglected. In particular, there is a chance that an intron is lost in such a pattern that its origin will be placed at a more recent inner node in the tree. For example, if an intron first appears in the MRCA for Ecdysozoa (similar example can be constructed for any phylogeny), it is possible that it is lost in *D. melanogaster* and *A. gambiae* and is only present in *C. elegans*. Then Dollo parsimony puts the origin of that intron onto the edge leading to *C. elegans*. Conversely, if the intron is lost in *C. elegans*, then Dollo parsimony places its origin at the node for Diptera. All methods agree (cf. Table 1) that such events cannot be too rare because many introns are lost on the branches leading to the insects and the worm. Another case in point are the 197 introns that are unique to *S. pombe* (44% of its introns). Dollo parsimony concludes that they were gained on that branch, which is doubtful.

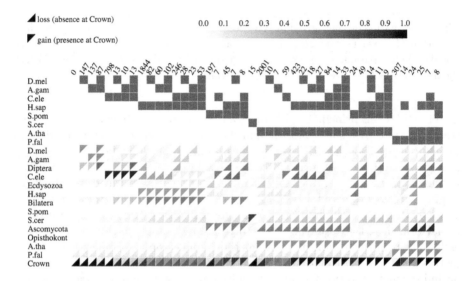

Fig. 4. Likelihoods for gains, losses, and presence at Crown for different characters. Columns correspond to characters: only those that occur at least seven times are shown. Character frequencies are displayed on top of the columns. Rectangles show the intron presence (shaded) or absence (empty) for each character. Shaded triangles show gain and loss posterior probabilities for each edge, and the posterior probabilities of intron presence/absence at the Crown taxon.

For characters that appear frequently in the data, Fig. 4 depicts probabilities for different scenarios. In some cases, the history is clear: if an intron is shared between *D. melanogaster* and *A. gambiae*, then there is a high probability of gain on the branch leading to Diptera, somewhat smaller one on the exterior branches leading to the two species, and some very small probabilities for gaining it earlier. In some cases, the posteriors show a mixture of possible histories: if an intron is present in *D. melanogaster* and *A. thaliana* (there are ten such cases), then it may have been gained more than once, or lost on several branches — which is not a surprising conclusion, but it illustrates the difficulty of choosing between such possibilities based on the intron presence/absence data alone. Notice also that the all-0 characters have no exciting history: most probably, they never had

Table 1. Intron evolution according to different methods. Values in the first row of each table are computed by Dollo parsimony, those in the second are computed by the formulas of Roy and Gilbert. The third row gives the posterior mean counts, computed via Eq. (2) assuming 35000 unobserved intron sites. The fourth row gives 95% confidence intervals for the posterior counts computed in a Monte Carlo procedure (see main text). Tree edges are identified by the nodes they lead to. Edges with a pronounced imbalance (at least 50%) towards gain or loss are emphasized in boldface.

(a) Intron counts at interior nodes

Method	Diptera	Ecdysozoa	Bilateria	Ascomycota	Opisthokont	Crown
Dollo parsimony (DP)	732	1081	1613	254	1046	978
Local likelihood (LL)	968	2305	3321	667	1903	1967
Posteriors (P)	895	1762	2380	554	1239	1064
P: 95% confidence	824–962	1484–1972	2055–2669	108–880	965–1450	692–1333

(b) Intron gains and losses on external branches

	D.mel.		A.gam.		C.ele.		H.sap.		S.pom.		S.cer.		A.tha.	
Method	gain	loss	gain	loss	gain	loss	gain	loss	gain	loss	gain	loss	gain	loss
DP	147	156	137	194	**798**	411	**1844**	112	**197**	1	15	**247**	**2001**	46
LL	90	**335**	91	**384**	719	**1555**	849	825	0	**167**	14	**656**	**1726**	760
P	116	**288**	111	**329**	855	1150	**1163**	200	0	**104**	15	546	**2157**	286
conf.	±27	±54	±24	±57	±46	±235	±239	±153	0	0–226	±3	102–871	±169	42–487

(c) Intron gains and losses on internal branches

	Diptera		Ecdysozoa		Bilateria		Ascomycota		Opisthokont	
Method	gain	loss	gain	loss	gain	loss	gain	loss	gain	loss
DP	87	**436**	36	**568**	**594**	27	3	**795**	92	24
LL	134	**1470**	0	**1005**	**1452**	35	308	**1536**	169	232
P	159	**1024**	141	752	**1216**	73	274	**953**	207	32
conf.	±60	187–618	0–256	±307	±286	0–151	0–553	±297	0–413	0–72

an intron present. Nevertheless, the small probabilities of gain and loss events associated with them add up to visible effects in the mean counts.

Table 1 compares three optimization criteria. Our estimates for intron counts, gains, and losses are mostly between the two previous estimates. Our likelihood-based approach gives only slightly more introns at the Crown than parsimony. The branch leading to *C. elegans* has more balanced gains and losses, which result in a net loss that is more modest than in [17]. The *H. sapiens* branch has almost six times as many gains than losses, as opposed to the likelihood calculations of [17] showing a balance. The branch leading to *A. thaliana* has a net gain predicted by all three methods. While we predict more gains on that branch than any of the other methods, the net change is close to what is computed by parsimony, due to more losses. Among the interior branches, we predict a significant net gain over the branch leading to the Opisthokont node, in agreement with parsimony, whereas [17] posit a modest net loss. As for pronounced biases towards gain or loss, our numbers agree with [17] concerning a tendency towards mass losses on a number of edges. At the same time, the mean counts tend to agree with parsimony regarding mass gains. In summary,

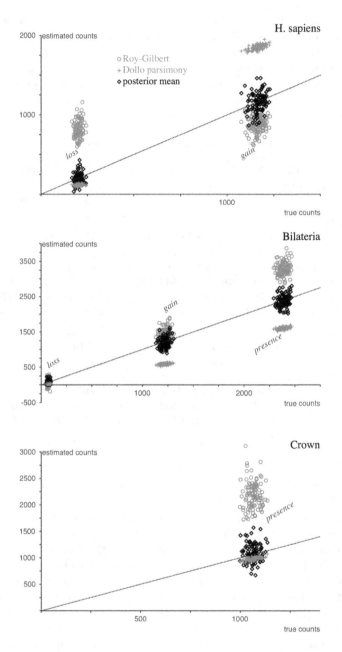

Fig. 5. Estimation error in 100 simulated data sets

our mean counts show more changes along the branches than parsimony, but are generally less extreme, and picture less intron-rich ancestral species than [17].

In order to assess the accuracy of the predictions in Table 1, we simulated intron evolution by the Markov model using the parameters optimized on the

Table 2. Rates of intron gain/loss on some external branches estimated by two methods: Roy and Gilbert [17] (RG) and our optimization. Rates for gains are given in units of 10^{-12}, while losses are in units of 10^{-9} per year. The branch lengths (same as in [17] to permit comparisons) are as follows: *D. melanogaster* 250–300 million years (MY), *C. elegans* 500–700 MY, *H. sapiens* 600–800 MY, *A. thaliana* 1500–2000 MY. The gain rates in the RG row are based on an assumption of as many intron sites as nucleotide positions: around 480 thousand, while our calculations are based on the assumption of 35000 unobserved intron sites. This difference amounts to a factor of about 12 between the two gain rate estimates: in parentheses we give numbers scaled to 480 thousand intron sites to permit direct comparison.

Method	D.mel. gain	D.mel. loss	C.ele. gain	C.ele. loss	H.sap. gain	H.sap. loss	A.tha. gain	A.tha. loss
RG	0.7–0.9	1.4–2.0	3.4–4.8	1.6–2.2	2.4–3.3	0.4–0.5	2.2–2.9	0.2–0.3
This paper	17–53	3.3–4.0	33–80	1.4–2.0	140–840	0.6–1.5	44–200	0.3–0.7
(scaled)	(1.4–4.5)		(2.8–6.9)		(11.8–72.8)		(3.8–17)	

original data set. The methods were applied to the simulated data sets to estimate intron counts, gains and losses, which could be compared to the exact values observed in the simulation. We generated 1000 synthetic data sets with the same number of observed intron sites in order to assess the estimation error of different methods in our probabilistic model.

Figure 5 plots the results of these experiments for some nodes. (For economy, only 100 experiments are shown: 1000 points would require a separate graph for every method at every node.) Our posterior counts generally perform better than the other two methods, which is not surprising in the case of Dollo parsimony (since its assumptions are decidedly different from those of our Markov model), but is more so for the formulas of Roy and Gilbert [16,17]. These latter usually underestimate intron gains and systematically overestimate the number of ancestral introns. It is also noteworthy that the formulas may sometime result in negative values, which need to be corrected to 0 manually. Dollo parsimony also tends to be biased against gains on internal edges but may overestimate them on external edges (Bilateria-*H. sapiens* edge in particular). It usually underestimates the number of ancestral introns. Aside from their bias, parsimony-based estimates have remarkably low variance. (In the simulations, the vector of ancestral intron counts is distributed multinomially with parameters depending on the likelihood of different characters. The same holds for the vector of intron gains or the vector of intron losses. The estimates of the other two methods have more complex distributions.) Our posterior counts do not seem to have any bias. For ancestral intron counts, the estimates deviate by at most a few hundred from their real values. Specifically, the number of ancestral introns at the common ancestor of animals, plants, and fungi is estimated with an error between (-372) and (+269) in 95% of the cases and a median error of (+11), whereas Dollo parsimony underestimates it by 85 on average (42–134 in 95% of the cases), and the formulas of [17] overestimate it by 1100 on average (710–1670 in 95% of the cases). The differences observed in the simulations are in fact very similar to those in Table 1.

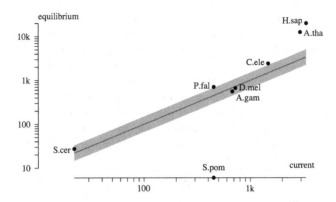

Fig. 6. Current and equilibrium intron counts. The latter are calculated from the stationary probabilities for the branch's Markov process. The dotted line shows identity, with a shaded band of ±50% around it. Species above the identity line are gaining introns, and species below it are losing introns.

Table 2 shows actual intron gain/loss rates calculated by optimization. Using the same actual time estimates for branch lengths as in [17], we computed the gain and loss rates in units of year^{-1}. Our ranges combine the uncertainty of branch lengths in years with 95% confidence intervals, calculated using the parametrized bootstrap procedure mentioned above, involving 1000 simulated data sets. Loss rates are comparable between previous and current estimates, but gain rates tend to be higher in our model. Most notably, gain rate on the branch leading from the MRCA of Bilateria to humans is by at least one magnitude higher than what was estimated in [17].

The Markov model enables predictions about intron dynamics in the future. Figure 6 compares current intron counts to the stationary probabilities for the appropriate branches: the Markov process on edge e converges to a ratio of $\mu_e : \lambda_e$ of intron absence:presence. *D. melanogaster*, *A. gambiae*, and *S. cerevisiae* are very close to equilibrium, but other organisms are farther from it. *C. elegans* is still within 50% of its stationary distribution, but *S. pombe* is losing introns, while humans and thale cress are heading toward much higher intron densities (six and four times as many introns as now, respectively).

5   Conclusion

We described probabilistic techniques for analyzing intron evolution, and applied them to a large data set. The probabilistic analysis assumes a Markov model of intron evolution, in which every intron site evolves independently, obeying the same rates, but the rates may be different on different branches. It is essential to allow for varying rates on branches because the mechanisms underlying intron gain and loss are fundamentally different, and their intensities vary between different organisms. We demonstrated that the model parameters can be estimated well from observing introns that evolved according to the model, and that the parameters provide sound estimates of ancestral intron counts. We described how posterior estimates can be computed exactly for ancestral intron counts and for gain and loss events. In contrast, Qiu *et al.* [5], relied on a reversible Markov model in which intron gain and loss rates are constant (for a particular gene family) across all branches of the tree. They further employed Markov chain Monte Carlo techniques to estimate posterior distributions.

Our analysis shows a dynamic history of introns, with frequent losses and gains in the course of eukaryotic evolution. We proposed a procedure for estimating unobserved intron sites. This procedure yields a more sound likelihood framework than what was used previously. Applied to the data set, which has 7236 orthologous intron sites, an additional 35000 unobserved intron sites were postulated to explain gains and losses. This equates to an intron site density of about one in every 12 nucleotides, which may characterize preferential intron insertion sites (such as exonic sequence motifs [5] enclosing the intron). All but 28 of 1064 introns present at the eukaryotic Crown node survived in at least one extant species, which means that about one seventh all introns predate the MRCA of animals, plants, and fungi, and the rest were gained more recently. Our counts show that about one third of human introns were gained after the split with Ecdysozoa, another third between that split and the split with fungi, and the rest mostly predate the MRCA of plants and animals.

It is conceivable that our model's assumptions of identical distribution and independence should be replaced by more realistic ones. We plan on exploring richer models in the future by enabling dependence between intron sites in the same gene, and by permitting varying rates among sites. Furthermore, by combining data analyzed here with new sequences, especially in light of recent analyses of introns in fungi [6] and nematoda [7], one can produce more nuanced results concerning intron evolution by better sampling the phylogenetic tree.

Acknowledgments. This research was supported by NSERC and FQRNT grants. I am grateful to Hervé Philippe, Scott Roy, and Igor Rogozin for valuable comments on this manuscript and on intron evolution in general. I would also like to thank the anonymous referees for their careful reading, and for suggesting the simulated experiments to assess the estimation error.

Supplemental information. The mentioned Java package and various additional analyses can be found at http://www.iro.umontreal.ca/~csuros/introns/.

References

1. Crick, W.: Split genes and RNA splicing. Science **204** (1979) 264–271
2. Lynch, M., Richardson, A.O.: The evolution of spliceosomal introns. Current Opinion in Genetics and Development **12** (2002) 701–710
3. de Souza, S.J.: The emergence of a synthetic theory of intron evolution. Genetica **118** (2003) 117–121
4. Rogozin, I.B., Wolf, Y.I., Sorokin, A.V., Mirkin, B.G., Koonin, E.V.: Remarkable interkingdom conservation of intron positions and massive, lineage-specific intron loss and gain in eukaryotic evolution. Current Biology **13** (2003) 1512–1517

5. Qiu, W.G., Schisler, N., Stoltzfus, A.: The evolutionary gain of spliceosomal introns: Sequence and phase preferences. Molecular Biology and Evolution **21** (2004) 1252–1263

6. Nielsen, C.B., Friendman, B., Birren, B., Burge, C.B., Galagan, J.E.: Patterns of intron gain and loss in fungi. PLoS Biology **2** (2004) e422

7. Coghlan, A., Wolfe, K.H.: Origins of recently gained introns in Caenorhabditis. Proceedings of the National Academy of Sciences of the USA **101** (2004) 11362–11367

8. Vaňáčová, Š., Yan, W., Carlton, J.M., Johnson, P.J.: Spliceosomal introns in the deep-branching eukaryote Trichomonas vaginalis. Proceedings of the National Academy of Sciences of the USA **102** (2005) 4430–4435

9. Ross, S.M.: Stochastic Processes. Second edn. Wiley & Sons (1996)

10. Rzhetsky, A., Ayala, F.J., Hsu, L.C., Chang, C., Yoshida, A.: Exon/intron structure of aldehyde dehydrogenase genes supports the "introns-late" theory. Proceedings of the National Academy of Sciences of the USA **94** (1997) 6820–6825

11. Rogozin, I.B., Lyons-Weiler, J., Koonin, E.W.: Intron sliding in conserved gene families. Trends in Genetics **16** (2000) 430–432

12. Felsenstein, J.: Inferring Pylogenies. Sinauer Associates, Sunderland, Mass. (2004)

13. Tuffley, C., Steel, M.: Links between maximum likelihood and maximum parsimony under a simple model of site substitution. Bulletin of Mathematical Biology **59** (1997) 581–607

14. Roch, S.: A short proof that phylogenetic reconstruction by maximum likelihood is hard. Technical report (2005) `math.PR/0504378` at `arXiv.org`.

15. Chor, B., Tuller, T.: Maximum likelihood of evolutionary trees is hard. In: Proc. Ninth Annual International Conference on Research in Computational Biology (RECOMB). (2005) In press.

16. Roy, S.W., Gilbert, W.: Complex early genes. Proceedings of the National Academy of Sciences of the USA **102** (2005) 1986–1991

17. Roy, S.W., Gilbert, W.: Rates of intron loss and gain: Implications for early eukaryotic evolution. Proceedings of the National Academy of Sciences of the USA **102** (2005) 5773–5778

18. Sverdlov, A.V., Rogozin, I.B., Babenko, V.N., Koonin, E.V.: Conservation versus parallel gains in intron evolution. Nucleic Acids Research **33** (2005) 1741–1748

19. Koshi, J.M., Goldstein, R.A.: Probabilistic reconstruction of ancestral protein sequences. Journal of Molecular Evolution **42** (1996) 313–320

20. Wolf, Y.I., Rogozin, I.B., Koonin, E.V.: Coelomata and not Ecdysozoa: Evidence from genome-wide phylogenetic analysis. Genome Research **14** (2004) 29–36

21. Roy, S.W., Gilbert, W.: Resolution of a deep animal divergence by the pattern of intron conservation. Proceedings of the National Academy of Sciences of the USA **102** (2005) 4403–4408

22. Philippe, H., Lartillot, N., Brinkmann, H.: Multigene analyses of bilaterian animals corroborate the monophyly of Ecdysozoa, Lophotrochozoa, and Protostomia. Molecular Biology and Evolution **22** (2005) 1246–1253

23. Philip, G.K., Creevey, C.J., McInerney, J.O.: The Opisthokonta and the Ecdysozoa may not be clades: Stronger support for the grouping of plant and animal than for animal and fungi and stronger support for the Coelomata than Ecdysozoa. Molecular Biology and Evolution **22** (2005) 1175–1184

24. Le Quesne, W.J.: The uniquely evolved character concept and its cladistic application. Systematic Zoology **23** (1974) 513–517

25. Farris, J.S.: Phylogenetic analysis under Dollo's law. Systematic Zoology **26** (1977) 77–88

OMA, A Comprehensive, Automated Project for the Identification of Orthologs from Complete Genome Data: Introduction and First Achievements

Christophe Dessimoz⋆, Gina Cannarozzi, Manuel Gil, Daniel Margadant, Alexander Roth, Adrian Schneider, and Gaston H. Gonnet

ETH Zurich, Institute of Computational Science, CH-8092 Zürich
cdessimoz@inf.ethz.ch

Abstract. The OMA project is a large-scale effort to identify groups of orthologs from complete genome data, currently 150 species. The algorithm relies solely on protein sequence information and does not require any human supervision. It has several original features, in particular a verification step that detects paralogs and prevents them from being clustered together. Consistency checks and verification are performed throughout the process. The resulting groups, whenever a comparison could be made, are highly consistent both with EC assignments, and with assignments from the manually curated database HAMAP. A highly accurate set of orthologous sequences constitutes the basis for several other investigations, including phylogenetic analysis and protein classification.

Complete genomes give scientists a valuable resource to assign functions to sequences and to analyze their evolutionary history. These analyses rely heavily on gene comparison through sequence alignment algorithms that output the level of similarity, which gives an indication of homology. When homologous sequences are of interest, care must often be taken to distinguish between orthologous and paralogous proteins [1].

Both orthologs and paralogs come from the same ancestral sequence, and therefore are homologous, but they differ in the way they arise: paralogous sequences are the product of gene duplication, while orthologous sequences are the product of speciation. Practically, the distinction is very useful, because as opposed to paralogs, orthologs often carry the same function, in different organisms. As Eugene Koonin states it [2], whenever we speak of "the same gene in different species", we actually mean orthologs.

1   Previous Large-Scale Efforts

The systematic identification of orthologous sequences is an important problem that several other projects have addressed so far. Among them, the COG

⋆ Corresponding author.

A. McLysaght et al. (Eds.): RECOMB 2005 Ws on Comparative Genomics, LNBI 3678, pp. 61–72, 2005.

database [3], [4] is probably the most established. From BLAST alignments [5] between all proteins ("all-against-all"), they identify genome-specific best hits, then group members that form triangles of best hits. Finally, the results are reviewed and corrected manually.

A further initiative is KEGG Orthology (KO) [6], [7]. KEGG is best known for its detailed database on metabolic pathways, but as the project evolved, an effort to cluster proteins into orthologous groups was initiated as well. The method is somewhat similar to COG: it starts with Smith-Waterman [8] all-against-all alignments, and identifies symmetrical best hits. It then uses a quasi-clique algorithm to generate "Ortholog clusters", that are used to create the KO groups, the last step being performed manually.

Finally, we mention here Inparanoid [9], OrthoMCL [10] and EGO (previously called TOGA) [11]. All three projects exclusively cover eukaryotic genomes. The two first insist on the inclusion of so-called "in-paralogs", sequences that result from a duplication event that occurred after all speciations. A noticeable short-coming of Inparanoid is the fact that it only handles pairs of genomes at a time. As for EGO, although their last release contains almost half a million genes from 82 eukaryotes, many sequences appear in more than one group and many groups contain paralogs. Because of that, we consider Inparanoid and EGO outside the present scope and limit our comparisons below to COG, KO and OrthoMCL.

2   Overview of the OMA Project

The project presented in this article is a new approach to identify groups of orthologs. It has some very specific properties:

- *Automated.* Unlike COG and KEGG Orthology, the whole workflow does not require human intervention, thereby insuring consistency, scalability and full transparency of the process.
- *Extensive.* The analysis so far has been performed on more than 150 genomes (Prokaryotes and Eukaryotes), with new ones added by the day[1]. The goal is to include all available complete genomes.
- *Strict.* Consistency checks are performed throughout the workflow, particularly at the integration step of genomic data. The algorithm for the identification of orthologous proteins excludes paralogs. 98.3% of the groups we could test are made of *bona fide* orthologous proteins (Sect. 4.1).

The algorithm for the identification of orthologous groups relies solely on protein sequence alignments from complete genomes, and hence does *not* depend on previous knowledge in terms of phylogeny, synteny information or experimental data. It is described in detail in the next section.

From the orthologous groups, we build a two-dimensional matrix in which each row represents an orthologous group and each column represents a species.

---

[1] At the time the final version of this article is submitted, 181 genomes have been included in the analysis.

The applications of that matrix are numerous and fall beyond the scope of this article. However, a few are worth mentioning. The rows provide phyletic patterns of the orthologous groups and can be used for phylogenetic profiling [12]. Parsimony trees can be constructed from the matrix to give either a phylogenetic tree when built from the columns, or protein families when built from the rows. We believe that both trees are very valuable contributions, and they will be presented, among others, in separate articles. Also, a large set of orthologous sequences is a prerequisite for the construction of reliable phylogenetic distance trees.

3   Methods

The construction of the matrix is performed in four steps. In the first one, genomic data is retrieved, checked for consistency and integrated. The second step consists of Smith-Waterman [8] protein alignments between all proteins ("all-against-all") followed by the identification of stable pairs, essentially what is sometimes also referred to as "symmetrical best hits". In the third step, the algorithm verifies every stable pair to ensure that it represents an orthologous relationship, not a paralogous one. Finally, in the fourth step, groups of orthologous proteins are formed from cliques of verified stable pairs.

3.1   Genome Data Retrieval, Verification and Integration

Complete genomes with protein sequence information are retrieved from Ensemble [13] and GenBank [14] and checked for consistency, then imported into *Darwin* [15], our framework. The consistency verification is extensive, and includes comparison between DNA and amino acid sequence, check for presence of start and stop codon, removal of fragments shorter than 50 amino acids, removal of duplicated sequences (sequences with >99% identity), verification of the total number of entries with HAMAP [16] (or GenBank/Ensembl for eukaryotes), and comparison with sequences present in SwissProt [17]. In case of alternative splicing, only the largest set of non-overlapping splice variants is kept for further analysis.

3.2   All-Against-All

Every protein sequence is aligned pairwise with every other protein sequence from a different organism using full dynamic programming [8]. The alignments were performed with GCB PAM matrices [18], using, for each alignment above noise level, the matrix corresponding to the PAM distance that maximizes the score, in a maximum likelihood fashion [19]. Alignments with score below 198 (70 bits, which typically corresponds to an E-value around 1.3e-16) or with length below 60% of the smaller sequence are considered not significant, and are discarded. The use of BLAST [5] was evaluated, but in the present case, we considered the speed increase not sufficient to compensate the loss in sensitivity [20]. Note that this view is shared by the teams behind KEGG Orthology [7] and STRING [21].

From the alignments, stable pairs are identified. That is essentially the idea behind what COG, among others, call "symmetrical best hits", that is, a protein pair in two different organisms that have each other as best match. However, as opposed to them, we improve robustness by keeping matches that have scores not significantly lower than the best match. Concretely, a stable pair can be formed between two proteins in two different organisms if, in both directions, the score of the alignment is not less than 90% of the best match.

3.3   Stable Pairs Verification

At this point, most stable pairs are expected to link two orthologous proteins, because orthologs usually have a higher level of similarity than paralogs. However, in case the corresponding ortholog of a particular protein is missing in some species (e.g. the organism lost it during evolution), a stable pair might be formed between that protein and a paralogous sequence, thus linking two proteins that belong to different orthologous groups. Such instances can be detected through the comparison to a third species that carries orthologs to both proteins (Figs. 1, 2). Therefore, each stable pair is verified through an exhaustive search against every other genome for such a scenario, and stable pairs corresponding to paralogy are discarded (Fig. 4). A more formal description of this algorithm, with proofs and examples are part of a separate publication.

3.4   Group Construction from Cliques of Verified Stable Pairs

The last step consists of orthologous groups identification from all verified stable pairs. The problem can be seen as a graph where proteins are represented by vertices and stable pairs by edges. In such a graph, an orthologous group is expected to form a fully connected subgraph. Thus, the algorithm iteratively looks for the maximal clique, groups the corresponding proteins and removes them from the graph. It runs until no more verified stable pairs are left. Finding

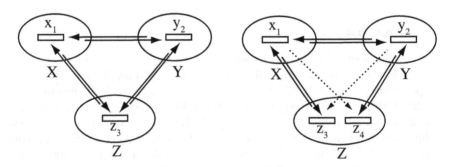

Fig. 1. Orthologous relationship between x_1 and y_2

Fig. 2. Paralogous relationship between x_1 and y_2, demonstrated by presence of z_3 and z_4 in Z

maximal cliques is a difficult problem (NP complete). The implementation of clique finding in *Darwin* [15] is based on the vertex cover problem and is a very effective clique approximation, which runs in reasonable time [22].

3.5 Tests for Accuracy and Completeness

On a project of such large size, it is crucial to ensure that all steps have been performed correctly, and that nothing is missing. With more than a hundred computers working around the clock for months, the probability of technical and operational failures becomes non-negligible, and must be proactively managed. We have included a number of tests that ensure quality all along the procedure described above. One test verifies that alignments are not missing through random sampling of 50,000 alignments per pair of genomes. Another test completely recomputes all recorded alignments of a pair of genomes, which is useful to detect (rare) errors due to hardware failure. A signature of the genomic database is computed at the end of each run to insure that memory was not corrupted during the computation. Yet another test verifies consistency of the results by looking for triangles of stable pairs that have a missing edge. More than once, these tests have revealed missing data, faulty hardware, and bugs in our programs.

4   Results and Discussion

The last OMA release classifies 501,636 proteins from 150 genomes into 111,574 orthologous groups (called *OMA groups* below). That covers 65.81% of all proteins contained in those genomes. The distribution of group size is such that most groups are small (Fig. 3). To a large extent, that is an obvious consequence of the large biodiversity among the included genomes. However, a technical reason can also explain part of that phenomenon: relatively few higher eukaryotes, and in particular plants, have been sequenced and included at this point, but they represent a significant portion of the total genes. All plant-specific genes in the matrix currently belong to groups of size two, simply because only two plants (*Arabidopsis thaliana* and *Oryza sativa*) are present. This effect is also reflected by a lower coverage of some eukaryotes. Therefore, we expect the group size and coverage to increase as more genomes are included.

The average group size is compared to other projects in Table 1. The differences are considerable. They can be explained by at least four factors: i) Quality of the algorithm. ii) Difference in the treatment of paralogous sequences. COG, KO, HAMAP and OrthoMCL often classify more than one protein per species into the same group. These proteins cannot have an orthologous relationship, by definition. In the best cases, those proteins are in-paralogs, genes that result from a duplication after all speciations, where justification for such inclusion is usually that in-paralogs are orthologous to all other proteins in the group. iii) Human validation. The practical problems of managing many groups are likely to create a bias toward fewer, larger groups (that can be observed in Table 1).

iv) Variation in the species composition and more generally in the biodiversity of the included sequences.

In that context, the size of the groups assigned by our algorithm does not appear unreasonable.

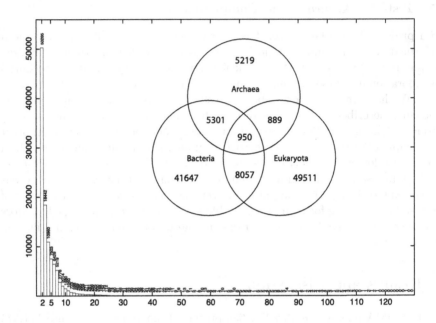

Fig. 3. Histogram of orthologous groups size and repartition of the 111,574 groups among kingdoms

Table 1. Comparison of some statistics accross projects. Note that KO and HAMAP only include partial genomes.

Project Name	Release	#Species	#Seqs	#Groups	Average Group Size	Coverage
COG	2003	66	138,458	4,873	28.4	75%
KO	22/Apr/2005	244	284,519	5,795	49.1	n/a
HAMAP	30/Apr/2005	876	26,977	1,071	25.2	n/a
OrthoMCL	I=1.5, 2003	7	47,668	7,265	6.6	47%
OMA	13/May/2005	150	501,636	111,574	4.5	66%

4.1   Validation

The quality of the groups resulting from our algorithm must be ensured. The statistics above about group size and genomes coverage constitute a first check, but more specific analysis of the results are desirable. This section presents the results from two further verifications, one using Enzyme Classification nomenclature, the other comparing our results with manual ortholog assignments from expert curators.

Function Validation Using Enzyme Classification. Enzyme Classification (EC) numbers are assigned based on the enzymatic activity of proteins. Since orthologs usually keep the same function, we expect in general that enzymes belonging to the same OMA group all have identical EC number. The Swiss Institute of Bioinformatics maintains the database [23] on Enzyme nomenclature that served us as reference (Release 37.0 of March 2005). First, the proteins that have more than one EC number (multi-functional enzymes, about 3% of all sequences in the EC database) were removed from the analysis. Then, every OMA group with at least two proteins that could be mapped to the EC database were selected for comparison.

There were 2,825 such groups out of 111,574 groups (2.5%). Of those, 2750 groups (97.3%) mapped to a single EC class. That compares very favorably to OrthoMCL, that has only 86% of its groups consistent with the EC assignments [10], although in their analysis, multi-functional enzymes were not excluded[2]. The result obtained for our method is particularly good if we consider that not all orthologs have identical function [24], and that the EC database is most probably not completely error-free.

Table 2. Comparison with HAMAP families

OMA Groups corresponding to HAMAP families:	1993	100%
— mapping to a single family:	1959	98.3%
— mapping to more than one family:	34	1.7%
HAMAP families corresponding to OMA Groups:	974	100%
— mapping to a single group:	355	36.4%
— mapping to more than one group:	619	63.6%

Comparison with HAMAP. Our groups were also compared with those of the HAMAP project [16]. As stated on their website, the HAMAP families are a collection of orthologous microbial proteins generated manually by expert curators. The comparison was done as following: in each HAMAP family, the in-paralogs were removed. OMA groups that had at least two proteins linkable to HAMAP were considered. Conversely, the HAMAP families with at least two proteins linkable to OMA groups were kept. Then, the correspondence between both sets of groups was assessed (Table 2). The results clearly show that while

---

[2] To compare the results with OrthoMCL in all fairness, the same analysis was performed on an OMA release from 26 eukaryotes, without removing multi-functional enzymes. With 1054 out of 1082 groups (97.4%) mapping to a single EC class, there was practically no difference.

our algorithm generates more/smaller groups than HAMAP, these groups are almost always consistent with the HAMAP assignments, in the sense that they each map to a single family. In fact, the numbers are such that slightly more than a third of the HAMAP families (347 out of 974) have a 1:1 correspondence to our groups, while most of the other two thirds are covered by typically two or three OMA groups. Considering that HAMAP families and OMA groups are constructed using radically different methodologies, this level of consistency is remarkable.

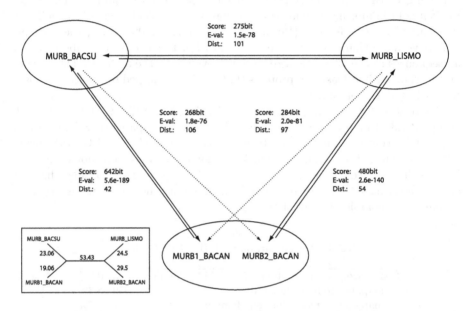

Fig. 4. Paralogs inside the HAMAP family MF_00037 (distances in PAM units)

The question that naturally arises from the comparison is whether it is our algorithm that has an excessive tendency to split orthologous groups or it is HAMAP that forms too large families. We performed some case-by-case analysis that revealed dubious classification on both sides: we have found several instances of OMA groups that have been split as a result of missing stable pairs (typically caused by alignment scores or length below our current threshold). Conversely, we found instances of sequences very likely to be paralogous within the same HAMAP family (Fig. 4). At this point, we are still investigating the relative merits of tighter versus larger groups.

4.2   Computational Cost

The all-against-all is the most time-consuming part of the computation. From the 150 genomes, we have 762,265 proteins producing about 2.85e11 pairwise alignments. In terms of the dynamic programming algorithm, the number of cells is 4.16e16 (where a cell corresponds to the computation of the table entry in the dynamic programming algorithm to align two sequences). We use *Darwin* [15] in parallel on more than 200 CPUs, which gives us a total capacity of about 1.9e13 cells/h (a Pentium IV at 3.2 GHz can perform about 2.2e10 cells/h). Hence, under such conditions, about 91 days would be required to compute the all-against-all. In practice, it took us longer, because of the consistency checks and changes in the data or programs. As for the stable pair verification and clique algorithm, they can be computed in about two days on a single machine.

4.3   Availability of the Results

The project homepage can be reached under `http://www.cbrg.ethz.ch/oma`. The list of species, progress of the all-against-all and OMA groups statistics are updated continuously. We offer a prototype online interface that enables users to browse through the results online.

5   Open Problems

As stated previously, the project is ongoing and some issues remain to be addressed. One of them concerns how to handle multi-domain proteins. The question is important, because the majority of proteins in Prokaryotes and Eukaryotes consist of at least two domains [25], where a domain is defined as an independent, evolutionary unit that can either form a single-domain protein or be part of a multi-domain one. Currently, our algorithm classifies a multi-domain protein with the group of its highest scoring domain. While that does not cause disruptive harm, it gives incomplete information about that multi-domain protein. In terms of consistency, it is not desirable to have that protein grouped with orthologs of its best scoring domain, while not grouped with orthologs of, say, its second best scoring domain. Either the focus is on domain orthology and it should be grouped to both, or the focus is on whole protein orthology and it should be grouped with none.

Lateral gene transfer is also a potential source of complications. Despite the abundant literature on the subject, the actual extent of this phenomenon remains unclear. Here as well, the effect on our group building process is non disruptive, xenologs are currently merely included in orthologous groups, but might cause problems in applications sensitive to phylogeny (e.g. phylogenetic trees). We are working on methods to systematically identify potential cases of lateral gene transfer *a posteriori*. The details and conclusions of this work will be the object of a separate publication.

6   Conclusion

The systematic identification of orthologuous sequences is an important problem in bioinformatics. In this article, we have presented OMA, a new large-scale project to cluster proteins into orthologous groups, where both the amount of data (150 genomes) and amount of computation (>500,000 CPU hours) justifies the large-scale description. Strict verification and consistency checks are performed throughout the workflow. The orthologous group construction is performed by an algorithm with several original features: it estimates a PAM distance between pairs of sequences matching significantly, it extends the concept of symmetrical best hit by considering all possible pairs of top matches within a tolerance factor, it detects and discards stable pairs connecting paralogous sequences and finally it identifies cliques of stable pairs to construct the groups. In contrast to most other projects, it does not rely on human validation. The resulting groups are highly consistent with EC assignments whenever applicable. They are also highly consistent with the manually curated database HAMAP, although our algorithm seems to have a tendency to split orthologous groups excessively. That issue, along with handling of multi-domain proteins and detection of lateral gene transfer events are the main problems that remain unsolved for now. However, even in its present state, we are confident that the project is an important contribution toward better identification of orthologous groups, and that it constitutes a solid basis for future work.

Acknowledgements

The authors thank Jean-Daniel Dessimoz, Markus Friberg and three anonymous reviewers for their comments and suggestions on the present manuscript, as well as Brigitte Boeckmann and Tania Lima from the Swiss Institute of Bioinformatics for useful discussions.

References

1. Fitch, W.M.: Distinguishing homologous from analogous proteins. Syst Zool **19** (1970) 99–113
2. Koonin, E.V.: An apology for orthologs - or brave new memes. Genome Biol **2** (2001) COMMENT1005
3. Tatusov, R.L., Koonin, E.V., Lipman, D.J.: A genomic perspective on protein families. Science **278** (1997) 631–7
4. Tatusov, R.L., Fedorova, N.D., Jackson, J.D., Jacobs, A.R., Kiryutin, B., Koonin, E.V., Krylov, D.M., Mazumder, R., Mekhedov, S.L., Nikolskaya, A.N., Rao, B.S., Smirnov, S., Sverdlov, A.V., Vasudevan, S., Wolf, Y.I., Yin, J.J., Natale, D.A.: The cog database: an updated version includes eukaryotes. BMC Bioinformatics **4** (2003) http://www.biomedcentral.com/1471–2105/4/41
5. Altschul, S.F., Madden, T.L., Schaffer, A.A., Zhang, J., Zhang, Z., Miller, W., Lipman, D.J.: Gapped BLAST and PSI-BLAST: a new generation of protein database search programs. Nucleic Acids Res **25** (1997) 3389–3402
6. Fujibuchi, W., Ogata, H., Matsuda, H., Kanehisa, M.: Automatic detection of conserved gene clusters in multiple genomes by graph comparison and P-quasi grouping. Nucleic Acids Res **28** (2000) 4029–4036

7. Kanehisa, M., Goto, S., Kawashima, S., Okuno, Y., Hattori, M.: The KEGG resource for deciphering the genome. Nucleic Acids Res **32** (2004) 277–280

8. Smith, T.F., Waterman, M.S.: Identification of common molecular subsequences. J. Mol. Biol. **147** (1981) 195–197

9. Remm, M., Storm, C., Sonnhammer, E.: Automatic clustering of orthologs and in-paralogs from pairwise species comparisons. J Mol Biol **314** (2001) 1041–52

10. Li, L., Stoeckert, C.J.J., Roos, D.S.: OrthoMCL: identification of ortholog groups for eukaryotic genomes. Genome Res **13** (2003) 2178–2189

11. Lee, Y., Sultana, R., Pertea, G., Cho, J., Karamycheva, S., Tsai, J., Parvizi, B., Cheung, F., Antonescu, V., White, J., Holt, I., Liang, F., Quackenbush, J.: Cross-referencing eukaryotic genomes: TIGR Orthologous Gene Alignments (TOGA). Genome Res **12** (2002) 493–502

12. Pellegrini, M., Marcotte, E.M., Thompson, M.J., Eisenberg, D., Yeates, T.O.: Assigning protein functions by comparative genome analysis: protein phylogenetic profiles. Proc Natl Acad Sci U S A **96** (1999) 4285–4288

13. Hubbard, T., Barker, D., Birney, E., Cameron, G., Chen, Y., Clark, L., Cox, T., Cuff, J., Curwen, V., Down, T., Durbin, R., Eyras, E., Gilbert, J., Hammond, M., Huminiecki, L., Kasprzyk, A., Lehvaslaiho, H., Lijnzaad, P., Melsopp, C., Mongin, E., Pettett, R., Pocock, M., Potter, S., Rust, A., Schmidt, E., Searle, S., Slater, G., Smith, J., Spooner, W., Stabenau, A., Stalker, J., Stupka, E., Ureta-Vidal, A., Vastrik, I., Clamp, M.: The Ensembl genome database project. Nucleic Acids Res **30** (2002) 38–41

14. Benson, D.A., Karsch-Mizrachi, I., Lipman, D.J., Ostell, J., Wheeler, D.L.: GenBank. Nucleic Acids Res **33 Database Issue** (2005) 34–38

15. Gonnet, G.H., Hallett, M.T., Korostensky, C., Bernardin, L.: Darwin v. 2.0 an interpreted computer language for the biosciences. Bioinformatics **16** (2000) 101–103

16. Gattiker, A., Michoud, K., Rivoire, C., Auchincloss, A.H., Coudert, E., Lima, T., Kersey, P., Pagni, M., Sigrist, C.J.A., Lachaize, C., Veuthey, A.L., Gasteiger, E., Bairoch, A.: Automated annotation of microbial proteomes in SWISS-PROT. Comput Biol Chem **27** (2003) 49–58

17. Boeckmann, B., Bairoch, A., Apweiler, R., Blatter, M.C., Estreicher, A., Gasteiger, E., Martin, M.J., Michoud, K., O'Donovan, C., Phan, I., Pilbout, S., Schneider, M.: The SWISS-PROT protein knowledgebase and its supplement TrEMBL in 2003. Nucleic Acids Res **31** (2003) 365–370

18. Gonnet, G.H., Cohen, M.A., Benner, S.A.: Exhaustive matching of the entire protein sequence database. Science **256** (1992) 1443–1445

19. Gonnet, G.H.: A tutorial introduction to computational biochemistry using Darwin. Technical report, Informatik, ETH Zurich, Switzerland (1994)

20. Brenner, S.E., Chothia, C., Hubbard, J.T.: Assessing sequence comparison methods with reliable structurally identified distant evolutionary relationships. Proc Natl Acad Sci U S A **95** (1998) 6073–6078

21. von Mering, C., Jensen, L.J., Snel, B., Hooper, S.D., Krupp, M., Foglierini, M., Jouffre, N., Huynen, M.A., Bork, P.: STRING: known and predicted protein-protein associations, integrated and transferred across organisms. Nucleic Acids Res **33 Database Issue** (2005) 433–437

22. Balasubramanian, R., Fellows, M.R., Raman, V.: An improved fixed-parameter algorithm for vertex cover. Inf. Process. Lett. **65** (1998) 163–168

23. Bairoch, A.: The ENZYME database in 2000. Nucleic Acids Res **28** (2000) 304–305

24. Jensen, R.A.: Orthologs and paralogs - we need to get it right. Genome Biol **2** (2001) INTERACTIONS1002
25. Vogel, C., Bashton, M., Kerrison, N.D., Chothia, C., Teichmann, S.A.: Structure, function and evolution of multidomain proteins. Curr Opin Struct Biol **14** (2004) 208–216

The Incompatible Desiderata of Gene Cluster Properties

Rose Hoberman[1] and Dannie Durand[2]

[1] Computer Science Department, Carnegie Mellon University, Pittsburgh, PA, USA
roseh@cs.cmu.edu
[2] Departments of Biological Sciences and Computer Science,
Carnegie Mellon University, Pittsburgh, PA, USA
durand@cmu.edu

Abstract. There is widespread interest in comparative genomics in determining if historically and/or functionally related genes are spatially clustered in the genome, and whether the same sets of genes reappear in clusters in two or more genomes. We formalize and analyze the desirable properties of gene clusters and cluster definitions. Through detailed analysis of two commonly applied types of cluster, r-windows and max-gap, we investigate the extent to which a single definition can embody all of these properties simultaneously. We show that many of the most important properties are difficult to satisfy within the same definition. We also examine whether one commonly assumed property, which we call *nestedness*, is satisfied by the structures present in real genomic data.

1 Introduction

Comparisons of the spatial arrangement of genes within a genome offer insight into a number of questions regarding how complex biological systems evolve and function. Spatial analyses of orthologous genomes focus on elucidating evolutionary processes and history, and on constructing comparative maps that facilitate the transfer of knowledge between organisms [1,2]. Conserved segments between different genomes have been used extensively to reconstruct the history of chromosomal rearrangements and infer an ancestral genetic map for a diverse group of species [3,4], as well as to provide novel features for new phylogenetic approaches. Genome self-comparisons reveal ancient large-scale or whole-genome duplication events [5]. Finally, spatial comparative genomics can also help predict protein function and regulation. In bacteria, conserved gene order and content have been used for prediction of operons, horizontal transfers, and more generally to help understand the relationship between spatial organization and functional selection [6–11].

A prerequisite to all of these tasks is the identification of genomic regions that share a common ancestor. Although offspring genomes immediately following speciation or a whole-genome duplication will have identical gene content and order, over time large and small scale rearrangements will obscure this relationship, leading to pairs of regions, or *gene clusters*, that share a number of homologous genes, but where neither order nor gene content is strictly conserved.

A. McLysaght et al. (Eds.): RECOMB 2005 Ws on Comparative Genomics, LNBI 3678, pp. 73–87, 2005.
© Springer-Verlag Berlin Heidelberg 2005

To identify such diverged homologous regions it is necessary to define the spatial patterns suggestive of common ancestry, and then design a search algorithm to find such patterns. The exact definition of the structures of interest is critical for sensitive detection of ancient homologies without inclusion of false positives. It is difficult to characterize what such regions will look like, however, since in most cases evolutionary histories are not known. Consequently, cluster definitions are generally based upon intuitive notions, derived either from small, well-studied examples (*e.g.,* such as the MHC region [12–14]), or from ideas about how rearrangements of genomes proceed. However, not much is known about the rates at which different evolutionary processes occur, and the little that is known is often based (somewhat circularly) on inferred homology of chromosomal segments.

The properties underlying existing cluster definitions are generally not stated, and the dimensions along which they differ have been analyzed in only a cursory manner. As a result, the formal tradeoffs between different models have been difficult to understand or compare in a rigorous way. Most cluster definitions are constructive, in the sense that they supply an algorithm to find clusters but do not specify explicit cluster criteria. In order to verify that an algorithm will identify all clusters satisfying the underlying intuitive criteria, however, these criteria must be stated formally. A few attempts have been made to formally define a gene cluster, but in these cases the focus tends to be on the design of an efficient and correct search algorithm, rather than on selecting a definition that captures those underlying intuitions. In addition to the cluster definition, the design of the search procedure may implicitly lead to additional unexpected or even undesirable properties, which would not be detected without explicit consideration of the cluster criteria. Finally, analysis of cluster properties can be useful for determining which characteristics actually reflect the types of structures found in real genomes, and thus which will best discriminate truly homologous regions from background noise (clusters of genes that occur by chance).

The goal of this paper is to characterize desirable properties of clusters and cluster definitions, in order to develop a more rigorous understanding of how modeling choices determine the types of clusters we are able to find, and how such choices influence the statistical power of tests of segmental homology. In Section 2, we describe the formal models and definitions discussed in this work. In Section 3, we present a set of properties upon which many existing gene cluster definitions, algorithms, and statistical tests are explicitly or implicitly based. We also propose additional properties that we believe are desirable, but are rarely stated explicitly. Through detailed analysis of two commonly applied types of cluster, r-windows and max-gap, we investigate the extent to which a single definition can embody all of these properties simultaneously. In Section 4, we examine whether one property that is implicitly assumed in many analyses, which we call *nestedness*, is actually satisfied by the structures present in real genomic data.

2   Models and Cluster Definitions

2.1   Models

We employ a commonly used model in which a genome is represented as an ordered set of n genes: $G = (g_1, \ldots, g_n)$. We assume a single unbroken chromosome, in which genes do not overlap. The distance between two genes in this model is simply the number of genes between them. In a *whole-genome comparison*, we are given two genomes G_1 and G_2, and a mapping between homologs in G_1 and G_2, where m of the genes in G_1 have homologs in G_2 (and vice versa). In this paper, we assume that each gene has at most one homolog in the other genome. We are interested in finding sets of homologs found in proximity in two different genomes (or possibly in two distinct regions of the same genome).

This model can be conceptualized in a number of ways, shown in Figure 1. Consider two genomes $G_1 = 1*2*34**56789$ and $G_2 = *3*14*2567*98$, where the integers correspond to homologous gene pairs, and the stars indicate genes with no homolog in the other genome. Figure 1(a) shows a *comparative map* representation, in which homologous pairs are connected by a line. Alternatively, in a *dot-plot* (Figure 1(b)), the horizontal axis represents G_1, the vertical axis represents G_2, and homologous pairs are represented as dots in the matrix. Finally, this data can be converted into an *undirected graph* (Figure 1(c)), where vertices correspond to homologous gene pairs. Two vertices are connected by

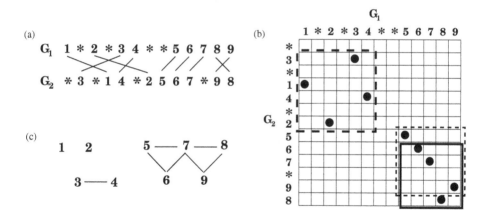

Fig. 1. Three ways in which to visualize a whole-genome comparison. Integers and stars denote genes, with stars denoting genes with no homolog in the other genome. (a) A comparative map. Lines show the mapping between homologous genes. (b) A dot plot showing the same information in a matrix format. Columns represent genes in G_1 and rows represent genes in G_2. A matrix element is filled with a black circle if the genes are homologous, and empty otherwise. (c) A graph in which vertices represent homologous gene pairs, and edges connect vertices if the corresponding genes are close together in both genomes. In this example, edges connect genes if the sum of the distances between the genes in both genomes is no greater than two.

an edge if the corresponding genes are close together in both genomes, where "close" is determined based on a user-defined distance function and threshold.

2.2   Cluster Definitions

A number of cluster definitions and algorithms have been proposed. In this paper we primarily focus on r-windows and max-gap clusters, two cluster definitions that are used in practice [6,7,15,16,17,9,10], but briefly describe other definitions as well.

An r-*window* cluster is defined as a pair of windows of r genes, one in each genome under consideration, in which at least k genes are shared [18,19,15]. This corresponds to a square in the dot-plot with sides of length r, which contains at least k homologs. For example, for a window model with $r = 5$ and $k = 4$, two clusters can be found in the example genome in Figure 1(b): {5,6,7,9} (dotted box) and {6,7,8,9} (solid box). We distinguish between the homologs shared in both instances of the cluster (the "marked" genes) and the intervening "unmarked" genes that occur in only one instance of the cluster (but which may have a homolog elsewhere in the genome).

The *max-gap* cluster definition also ignores gene order and allows insertions and deletions, but does not constrain the maximum length of the cluster to r genes. Instead, a max-gap cluster is described by a single parameter g, and is defined as a set of marked genes where the distance (or *gap*) between adjacent marked genes in each genome is never larger than a given distance threshold, g [20,21]. When $g = 0$, max-gap clusters are referred to as *common intervals* [22–24]. When the maximum gap allowed is $g = 1$, two maximal max-gap clusters are found in the example genome in Figure 1(b): {1,2,3,4} (dashed box) and {5,6,7,8,9} (not shown). A max-gap cluster is *maximal* if it is not contained within any larger max-gap cluster. Correct search algorithms for this definition require some sophistication. Bergeron *et al.* originally developed a divide-and-conquer algorithm to conduct a whole-genome comparison, and efficiently detect all maximal max-gap clusters [20]. Many groups design heuristics to find max-gap clusters, but such methods are not guaranteed to find all maximal max-gap clusters.

Other definitions include that of Calabrese *et al.* [25], in which the distance between each pair of homologs is evaluated as a function of the gap size in *both* genomes. Unlike the max-gap definition, which only requires that in both genomes the distance to *some* other marked gene in the cluster is small, this method requires that all marked genes that are adjacent in genome G_1 *also* be close in genome G_2, but not vice versa. A very different approach by Sankoff *et al.* [26] explicitly evaluates a cluster (or segment) by a weighted measure of three properties: compactness, density, and integrity. They seek a global partition of the genome into segments such that the sum of segment scores is minimized. Clusters have also been defined in terms of graph-theoretic structures (*e.g.*, Figure 1(c)), such as connected components [27] or high-scoring paths [28,29]. Finally, a variety of heuristics have been proposed to search for gene clusters [30,25,31,32,33,34,29,11], the majority of which are specifically designed to find sets of genes in approximately collinear order (*i.e.,* forming a rough diagonal on the dot-plot).

3   Cluster Properties

Many of the cluster properties underlying existing definitions derive from the processes that lead to genome rearrangements. As genomes diverge, large-scale rearrangements break apart homologous regions, reducing the size and length of clusters. Gene duplications and losses cause the gene complement of homologous regions to drift apart, so that many genes will not have a homolog in the other region, and gene clusters will appear less dense. Smaller rearrangements will disrupt the gene order and orientation within homologous regions. Thus, clusters are often characterized according to their size, length, density, and the extent to which order and orientation are conserved. We discuss these properties in more detail below, as well as a number of additional properties that are rarely stated explicitly, but that we argue are nonetheless desirable.

Size: Almost all methods to evaluate clusters consider the size of a cluster, *i.e.,* the number of marked genes contained within it. In general it is assumed that the more homologs in a cluster, the more likely it is to indicate common ancestry rather than chance similarities. An appropriate minimum size threshold will depend, however, on the specific cluster definition. For example, a cluster of four homologs in which order is conserved may be less likely to occur by chance, and thus more significant than an unordered cluster of size four.

Length: The length of a cluster, defined with respect to a particular genome, is the total number of marked and unmarked genes contained within it. For example, in Figure 1(b), the upper left cluster is of size four, and spans two unmarked genes, so is of total length six. In a whole-genome comparison, the number of unmarked genes spanned by the cluster in each genome may differ. However, if the processes that degrade a cluster are operating uniformly, then the length of the cluster in both genomes should be similar. This similarity of lengths is implicitly sought by the length constraint of r-windows, and explicitly sought in the clustering method of Hampson *et al.* [33].

Density: Although over time gene insertions and losses will cause the gene content of homologous regions to diverge, in most cases we expect that significant similarity in gene content will be preserved. Thus, the majority of existing approaches attempt to find regions that are densely populated with homologs. We define the *global density* of a cluster as its size divided by its length. For example, in Figure 1(b), the first max-gap cluster is of size four and length six, so has a density of $2/3$. For a fixed value of r, the minimum global density of an r-window is set by choosing the parameter k. The only way to set a constraint on the global density of a max-gap cluster, on the other hand, is to reduce g, which will also reduce the maximum length of a cluster.

Even when a minimum global density is required, regions of a cluster may not be locally dense: a cluster could be composed of two very dense regions separated by a large region with no homologs. In this case, it might seem more natural to break the cluster into two separate clusters. Density as we have defined it here reflects the average gap size, but does not reflect the *variance* in gap sizes. The gap between adjacent marked genes in an r-window can be as large as $r-k$, whereas max-gap clusters guarantee that the maximum gap will be no more than g. Note that the two definitions have switched roles: the local density is easily controlled by the parameter g for max-gap clusters but there is no way to constrain the local density of r-window clusters without also further constraining the maximum cluster length. This trade-off between global and local density gives a simple illustration of how it can be difficult to design a cluster definition that satisfies our basic intuitions about cluster properties.

Order: For whole-genome comparison, a cluster is considered ordered if the homologs in the second genome are in the identical or opposite order of the homologs in the first genome. For example, consider the two genomes shown in Figure 1(b). The clusters {5,6,7} and {8,9} are ordered, but {1,2,3,4} is not. Many cluster definitions require a strictly conserved gene order [6,31,11]. Over time, however, inversions will cause rearrangements, and thus conserved gene order is often considered too strict a requirement. In order to allow some short inversions, Hampson *et al.* [32] explicitly parameterize the number of order violations that are allowed in a cluster. A number of groups use heuristic, constructive methods that either implicitly enforce certain constraints on gene order, or explicitly bias their method to prefer clusters that form near-diagonals in the dot plot [25,34,29,17]. The remainder, including r-windows and max-gap clusters, completely disregard gene order. As we will see, however, though a number of groups *state* that they ignore gene order, constraints on gene order are often unintended consequences of algorithmic choices (see nestedness).

Orientation: Conserved spatial organization in bacterial genomes often points to functional associations between genes. In particular, clusters of genes in close proximity, with the same orientation, often indicate operons. In whole-genome comparison of eukaryotes, similarities in gene orientation can provide additional evidence that two regions share a common ancestor. To the best of our knowledge, however, except for the method of Vision *et al.* [29], in which changes in orientation decrease the cluster score, existing definitions either require all genes in a cluster to have the same orientation, or disregard orientation altogether.

Temporal Coherence: Temporal information can be used to evaluate the significance of a putative homologous region identified through whole-genome comparison. If a set of homologous genes all arose through the same speciation or duplication event, then the points in time at which each homolog pair diverged will be identical, and consequently we would expect our estimates of these divergence times to be similar. However, all existing methods to find clusters are based solely on spatial information, and divergence times have been used only to estimate the age of a duplicated block identified based on spatial organization [6,35], but not to assess the statistical significance of a cluster. In theory, combined analysis of temporal and spatial information could be used, for example, to increase our confidence that a region is the result of a single large-scale duplication event. However, due to the large error bounds that must be associated with any sequence-based estimate of divergence times [36,37,38], the practicality of such an approach is as yet unclear.

Nestedness: For whole-genome comparison, one cluster property that is generally not considered explicitly, but may be assumed implicitly, is nestedness. A cluster of size k is *nested* if for each $h \in 1 \ldots k - 1$ it contains a valid cluster of size h. Intuitively it may seem that any reasonable cluster definition should have this property. In fact, clusters with no ordering constraints are not necessarily nested. For example, Bergeron *et al.* [20] state a formal definition of max-gap clusters, and prove that there are maximal max-gap clusters of size k which do not contain any valid sub-cluster of size $2..k-1$. For example, when $g = 0$ they present a non-nested max-gap cluster with only four genes. The sequence of genes 1234 on one genome and 3142 on the other form a max-gap cluster of size four which does not contain any max-gap cluster of size two or three. Thus, nested max-gap clusters comprise only a subset of general max-gap clusters found through whole-genome comparison.

There are no definitions that explicitly require that clusters be nested; rather, greedy search algorithms implicitly limit the results to nested clusters. Greedy algorithms use a bottom-up approach: each homologous gene pair serves as a cluster seed, and a cluster is extended by looking in its chromosomal neighborhood for another homologous gene pair close to the cluster on both genomes [25,31,33,39]. It can be shown that any greedy search algorithm that constructs max-gap clusters iteratively, *i.e.*, by constructing a cluster of size k by adding a gene to a cluster of size $k - 1$, will find *exactly* the set of all maximal nested max-gap clusters, as long as it considers each homologous gene pair as a seed for a potential cluster. In such cases, although order is not explicitly constrained, the search algorithm enforces implicit constraints on gene order: nested clusters can only get disordered to a limited degree. In most cases, however, such constraints are not acknowledged, and perhaps not even recognized.

Disjointness: If two clusters are not disjoint, *i.e.,* the intersection of the marked genes they contain is not empty[1], our intuitive notion of a cluster may correspond more closely to the single island of overlapping windows than to the individual clusters. For example, Figure 1(b) shows two windows for which $r = 5$ and $k = 4$: {5,6,7,9} and {6,7,8,9}. Although both clusters contain genes 6, 7, and 9, there is no window of length five that contains all five of the genes. Thus, r-windows are not always disjoint. Indeed, it is surprisingly hard to find a cluster definition that guarantees that all clusters will be disjoint. The majority of definitions lead to overlapping clusters that must be merged or separated in an ad-hoc post-processing step for use by algorithms that require a unique tiling of regions. The only definition for which maximal clusters have been shown to be disjoint

---

[1] Note that it is possible, however, for two disjoint clusters to have overlapping spans in one of the genomes, as long as they do not share any homologs.

is the max-gap cluster [20]. However, when adding additional constraints in addition to the maximum gap size, disjointness is quickly forfeited. For example, consider the consequences of requiring conserved order when looking for max-gap clusters in Figure 1(a). With a maximum gap of $g = 2$, three clusters with conserved order are identified $\{1,2\}$, $\{3,4,5,6,7,8\}$, $\{3,4,5,6,7,9\}$. Although the last two clusters overlap, they cannot be merged without breaking the ordering constraint (due to the inversion of the segment containing genes 8 and 9).

More generally, a lack of disjointness strongly suggests that the cluster definition is too constrained. In the r-window example, these clusters are not disjoint *precisely* because the definition artificially constrains the length of a cluster. In the second example, the clusters were not disjoint because a definition with a strict ordering constraint was not able to capture the types of processes, such as inversions, that created the cluster.

Isolation: If we observe a cluster with some additional homologous pairs in close proximity to its borders we might feel that the cluster border was arbitrary, and should extend to cover the neighboring island of genes. Thus, we propose that cluster definitions should guarantee that clusters will be *isolated*, that is: the maximum distance between marked genes in a cluster should always be less than the minimum distance between two clusters. A maximum-gap constraint guarantees that clusters will be isolated, but only barely—the gap within a cluster may be as large as g, whereas the gap separating two clusters may be just $g+1$.

Symmetry: For whole-genome comparison, a desirable property that is rarely considered explicitly is whether the definition is symmetric with respect to genome. In some cases, such as the definition proposed by Calabrese *et al.* [25], a cluster is defined in such a way that whether a set of genes form a valid cluster may depend on whether genome G_1 or genome G_2 is represented by the vertical axis in the dot-plot. Put another way, the set of clusters identified will differ depending on which genome is designated as the reference genome. A surprisingly large proportion of constructive definitions are not symmetric. These clustering algorithms require the selection of a reference genome even when there is no clear biological motivation for this choice. Definitions that are symmetric with respect to genome include r-windows and max-gap cluster definitions, as well as algorithms that represent the dot-plot as a graph and use a symmetric distance function [27,29].

4   Are Max-Gap Clusters in Genomic Data Nested?

Cluster definitions that constrain the gap size between marked genes are widely used in genomic studies [30,6,40,7,16,17,9,41,10,34,29]. In the majority of cases, however, clusters are detected with a greedy algorithm, whereby larger clusters are identified by extending smaller clusters. Remember that greedy methods find the subset of max-gap clusters that are nested and that nestedness implies a certain degree of ordering. It is not clear whether greedy methods are used for

Table 1. The genomes compared (G_1 and G_2), the total number of genes in each genome (n_1 and n_2, respectively), and the number of orthologs identified, excluding ambiguous orthologs (m)

G_1	G_2	n_1	n_2	m
E. coli	B. subtilis	4, 108	4, 245	1, 315
Human	Mouse	22, 216	25, 383	14, 768
Human	Chicken	22, 216	17, 709	10, 338

```
for i= 1..n do      // i iterates through all genes in G₁
  C = {i};          // C is the cluster being constructed
  L₁ = R₁ = i;      // Lᵢ and Rᵢ are the left/rightmost positions in C on Gᵢ
  L₂ = R₂ = p(i);   // p(i) indicates the position of gene i's homolog in G₂
  j = L₁-g-1;       // j iterates through all genes close to C on G₁
  while (L₁-g-1 ≤ j ≤ R₁+g+1) do
      if j ∉ C and p(j) ∈ {L₂-g-1, ..., R₂+g+1} // if j is close to C in G₂
        C = C ∪ j;                              //    add it to C
        L₁ = min(L₁,j);   L₂ = min(L₂,p(j));
        R₁ = max(R₁,j);   R₂ = max(R₂,p(j));
        j = L₁-g-1;                             // start the search over
      else
        j++;
      end
    end
  clusters = clusters ∪ C;
end
```

Fig. 2. Pseudo-code for a greedy, bottom-up algorithm to find nested max-gap clusters

computational convenience or because researchers believe that nested clusters better capture the biological processes of interest. In this section, we investigate the practical consequences of choosing one search procedure over the other. We compare three pairs of genomes to determine the proportion of max-gap clusters in real genomes that are actually nested.

Whole-genome comparisons of three pairs of genomes at varying evolutionary distances were conducted. The first comparison was of *E. coli* and *B. subtilis*, with a mapping of orthologs between the two genomes obtained from the GOLDIE database [30]. The other two comparisons were of human and mouse, and human and chicken, with ortholog mappings obtained from the InParanoid database [42]. The total number of genes in each genome, and the number of orthologs identified, is given in Table 1.

The GeneTeams software, an implementation of the top-down algorithm of Bergeron *et al.* [20], was used to identify all maximal max-gap clusters shared between the two genomes, for $g \in \{1, 5, 10, 15, 20, 30, 50\}$. In addition, we designed a simple bottom-up, greedy algorithm to identify all maximal *nested* max-gap clusters (Figure 2). This algorithm considers each pair of orthologs in turn, treating each as a cluster seed from which a greedy search for additional orthologs

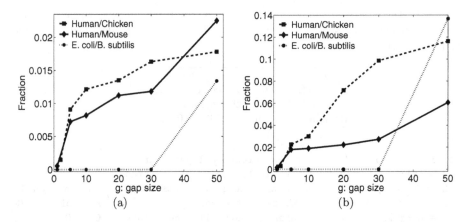

Fig. 3. Comparison of the set of nested clusters to the set of gene teams, for $g \in \{1, 5, 10, 15, 20, 30, 50\}$. (a) The fraction of gene teams that are *not* nested. (b) The fraction of maximal nested clusters that are *not* gene teams.

is initiated. Occasionally different seeds may yield identical clusters. Any such duplicate clusters are filtered out, as are non-maximal nested clusters (clusters strictly contained within another nested cluster). However, overlapping clusters (*e.g.,* properly intersecting sets) are not merged together, since the resulting merged clusters would not be nested.[2]

For the bacterial comparison, for all gap values except $g = 50$, both methods found the same set of clusters, *i.e.,* all gene teams were nested. In all eukaryotic comparisons, however, at least one non-nested gene team was identified. Nonetheless, the percentage of teams that were not nested remained low for all comparisons, ranging from close to 0% to about 2% as the gap size was increased (Figure 3(a)). The percentage of nested clusters that were not gene teams (in other words, clusters that could have been extended further if a greedy algorithm had not been used), was also close to zero for small gap sizes, but increased more quickly, peaking at almost 15% for a gap size of $g = 50$ (Figure 3(b)). In contrast, in randomly ordered genomes, although large gene-teams are much rarer, a much higher percentage are not nested (data not shown).

Another quantity of interest is the number of *genes* that would be missed altogether if a greedy approach is used rather than a top-down algorithm; that is, the number of genes that are found in a large gene team but not in a large nested cluster. For a minimum cluster size of two, very few genes are missed: the number of genes missed remains under 20 for both eukaryotic datasets, no matter how large the gap size (Figure 4, circles). For a more realistic minimum cluster size of seven, however, the number of missed genes rises more quickly,

---

[2] It is unclear whether those who employ a greedy heuristic merge all overlapping clusters or not, since such heuristics are generally specified quite vaguely, if at all. However, in our datasets, only a small percentage of clusters detected with the greedy algorithm overlapped (*e.g.,* 2% in the human/chicken comparison).

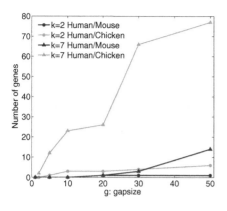 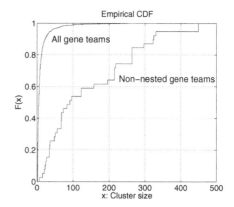

Fig. 4. The number of genes in a gene team of size $k \geq 2$, that are not in *any* nested max-gap cluster of size $k \geq 2$ (circles). The triangles show the number of genes when the minimum cluster size is seven.

Fig. 5. A CDF comparing the distribution of gene team sizes to the distribution of nested gene team sizes, for human vs chicken, for all gap sizes tested

peaking near 80 for the human/chicken comparison (Figure 4, triangles), and near 120 for the bacterial comparison (data not shown).

The gene teams that are not nested tend to be the larger clusters. For example, Figure 5 compares the distribution of gene teams sizes to the distribution of non-nested gene teams sizes, for the human/chicken comparison, for the complete set of clusters identified at any gap size. The gene team size distribution peaks very quickly: over 80% of gene teams contain fewer than ten genes. The sizes of non-nested gene teams, however, peak much more slowly: only about 10% of non-nested gene teams contain fewer than ten genes. It is not until the size reaches 270 genes that the CDF reaches 0.8.

In summary, when comparing *E. coli* with *B. subtilis* with reasonable gap sizes, the nestedness assumption does not exclude any clusters from the data. For the eukaryotic datasets, these results also suggest that for smaller gap sizes few clusters are missed when using a greedy search strategy. For larger gap values, the nestedness assumption does appear to lead to some loss of signal, especially in the human/chicken comparison: large clusters are identified only in fragments, and the spatial clustering of many genes is not detected at all. For more diverged genome pairs, as clusters become more disordered, this loss of signal may be exacerbated. This remains to be investigated, as do the practical implications of the nestedness assumption on the detection of duplicated segments through genome self-comparison.

5   Discussion

We have characterized desirable properties of cluster definitions, and compared a number of existing definitions with respect to these properties. The detailed catalog of cluster properties presented here will be useful for assessing whether definitions satisfy the intuitive notions upon which they are implicitly based, and whether these notions actually correspond to the types of structures present in real-genomic data. Analyses of desirable cluster properties may also pave the way for new, possibly more powerful cluster definitions.

Our analysis of cluster properties reveals that existing approaches to identifying gene clusters differ both in terms of the characteristics of the clusters they were explicitly designed to find, and in terms of the properties that emerge as unintended consequences of modeling choices. We show that the search procedure, in addition to the cluster definition, often implicitly enforces additional types of constraints. Such implicit constraints may be particularly problematic when the goal is to characterize the properties of homologous regions. For example, although the CloseUp algorithm was ostensibly designed to identify chromosomal homology using "shared-gene density alone" [33], the greedy nature of the search algorithm means that all clusters with a minimum gene density may not actually be detected. If such an approach was used to evaluate the extent to which order is conserved in homologous regions, incorrect inferences could be made. For example, if clusters with highly scrambled gene order were not found, one might erroneously conclude that no such clusters exist, rather than that the clustering algorithm was simply not capable of finding them. Without a clear understanding of which properties are constrained by the method, and which properties are inherent in the data, it can be difficult to interpret such results.

Our results also show that, for the datasets considered here, a greedy search strategy for max-gap clusters may actually improve statistical power, at least for small gap sizes. A test of cluster significance will have increased power (*i.e.*, a reduced number of false negatives) when the cluster definition is as narrow as possible, while still capturing the properties exhibited by diverged homologous regions. These properties, however, are generally not known, since there is little data about evolutionary histories or processes. In some cases, however, the appropriateness of a particular property can be evaluated even without full knowledge of evolutionary histories. For example, if adding an additional constraint to the cluster definition does not eliminate any of the clusters identified in the data, then we argue that it is not only acceptable to include such a property in the cluster definition, but desirable, in order to increase statistical power. Thus, when comparing *E. coli* with *B. subtilis* with reasonable gap sizes, a nested cluster definition appears to be a good choice: the nestedness assumption does not exclude any clusters from the data, but significantly reduces the probability of observing a cluster by chance, thereby strengthening the measurable significance of detected clusters.

These results also suggest that in the three datasets we studied most clusters remain quite ordered. Although an assumption of nestedness does implicitly constrain gene order, more quantitative measures of order conservation may be found that increase statistical power still further. How to best quantify the degree to which order is conserved, however, remains an open question.

Although there is often overlap among the properties of different definitions, there is as yet no consensus on what criteria best reflect biologically important features of gene clusters. This lack of consensus reflects the sparsity of data about evolutionary histories and evolutionary processes, and also that the relevance of particular properties depends to a large degree on the dataset being analyzed, as well as the researcher's goals. For example, physical distances between genes and gene orientation may not be very helpful for identifying homology between eukaryotic genomes, but may be important for identifying functional clusters in bacteria. For identifying gene duplications, which are often followed by significant differential gene loss of the homologs on each duplicated segment [43], gene density may be of reduced importance than for identifying paralogous segments. In addition, when clusters are being identified as a pre-processing step for reconstructing rearrangement histories, the exact boundaries and sizes of the cluster may be quite important [44]. In other cases, a researcher may be trying to test a global hypothesis (such as finding evidence for one or two rounds of whole-genome duplication), and may not necessarily care about the significance or boundaries of any specific cluster.

Even if it were known which properties reflect biologically relevant features, designing a definition to satisfy those properties may not be straightforward because, in many cases, properties are not independent. Properties may interact in subtle ways—a definition that guarantees one desirable property will often fail to satisfy another. For example, one of the nice properties of the max-gap definition is that clusters are always disjoint. However, as shown in Section 3, adding additional constraints on order or length results in clusters that are no longer guaranteed to be disjoint. The subtle and sometimes undesirable interplay of some of these properties makes it difficult to devise a definition that satisfies them all. In fact, many of the most important properties are difficult to satisfy with the same definition. Thus, it remains an open question to what extent a single definition can capture all of these properties simultaneously.

Acknowledgment

D.D. was supported by NIH grant 1 K22 HG 02451-01 and a David and Lucille Packard Foundation fellowship. R.H. was supported in part by a Barbara Lazarus Women@IT Fellowship, funded in part by the Alfred P. Sloan Foundation. We thank B. Vernot and N. Raghupathy for comments on the manuscript, and David Sankoff for helpful discussion and for suggesting the title of the paper.

References

1. Murphy, W.J., Pevzner, P.A., O'Brien, S.J.: Mammalian phylogenomics comes of age. Trends Genet **20** (2004) 631–9
2. O'Brien, S.J., Menotti-Raymond, M., Murphy, W.J., Nash, W.G., Wienberg, J., Stanyon, R., Copeland, N.G., Jenkins, N.A., Womack, J.E., Graves, J.A.M.: The promise of comparative genomics in mammals. Science **286** (1999) 458–81

3. Sankoff, D.: Rearrangements and chromosomal evolution. Curr Opin Genet Dev **13** (2003) 583–7

4. Sankoff, D., Nadeau, J.H.: Chromosome rearrangements in evolution: From gene order to genome sequence and back. PNAS **100** (2003) 11188–9

5. Simillion, C., Vandepoele, K., de Peer, Y.V.: Recent developments in computational approaches for uncovering genomic homology. Bioessays **26** (2004) 1225–35

6. Blanc, G., Hokamp, K., Wolfe, K.H.: A recent polyploidy superimposed on older large-scale duplications in the *Arabidopsis* genome. Genome Res **13** (2003) 137–144

7. Chen, X., Su, Z., Dam, P., Palenik, B., Xu, Y., Jiang, T.: Operon prediction by comparative genomics: an application to the Synechococcus sp. WH8102 genome. Nucleic Acids Res **32** (2004) 2147–2157

8. Lawrence, J., Roth, J.R.: Selfish operons: horizontal transfer may drive the evolution of gene clusters. Genetics **143** (1996) 1843–60

9. Overbeek, R., Fonstein, M., D'Souza, M., Pusch, G.D., Maltsev, N.: The use of gene clusters to infer functional coupling. Proc Natl Acad Sci U S A **96** (1999) 2896–2901

10. Tamames, J.: Evolution of gene order conservation in prokaryotes. Genome Biol **6** (2001) 0020.1–11

11. Wolf, Y.I., Rogozin, I.B., Kondrashov, A.S., Koonin, E.V.: Genome alignment, evolution of prokaryotic genome organization, and prediction of gene function using genomic context. Genome Res **11** (2001) 356–72

12. Endo, T., Imanishi, T., Gojobori, T., Inoko, H.: Evolutionary significance of intragenome duplications on human chromosomes. Gene **205** (1997) 19–27

13. Smith, N.G.C., Knight, R., Hurst, L.D.: Vertebrate genome evolution: a slow shuffle or a big bang. BioEssays **21** (1999) 697–703

14. Trachtulec, Z., Forejt, J.: Synteny of orthologous genes conserved in mammals, snake, fly, nematode, and fission yeast. Mamm Genome **3** (2001) 227–231

15. Friedman, R., Hughes, A.L.: Gene duplication and the structure of eukaryotic genomes. Genome Res **11** (2001) 373–81

16. Luc, N., Risler, J., Bergeron, A., Raffinot, M.: Gene teams: a new formalization of gene clusters for comparative genomics. Comput Biol Chem **27** (2003) 59–67

17. McLysaght, A., Hokamp, K., Wolfe, K.H.: Extensive genomic duplication during early chordate evolution. Nat Genet **31** (2002) 200–204

18. Cavalcanti, A.R.O., Ferreira, R., Gu, Z., Li, W.H.: Patterns of gene duplication in *Saccharomyces cerevisiae* and *Caenorhabditis elegans*. J Mol Evol **56** (2003) 28–37

19. Durand, D., Sankoff, D.: Tests for gene clustering. Journal of Computational Biology (2003) 453–482

20. Bergeron, A., Corteel, S., Raffinot, M.: The algorithmic of gene teams. In Gusfield, D., Guigo, R., eds.: WABI. Volume 2452 of Lecture Notes in Computer Science. (2002) 464–476

21. Hoberman, R., Sankoff, D., Durand, D.: The statistical significance of max-gap clusters. In Lagergren, J., ed.: Proceedings of the RECOMB Satellite Workshop on Comparative Genomics, Bertinoro, Lecture Notes in Bioinformatics, Springer Verlag (2004)

22. Didier, G.: Common intervals of two sequences. In: WABI. Volume 2812., Lecture Notes in Computer Science (2003) 17–24

23. Heber, S., Stoye, J.: Algorithms for finding gene clusters. In: WABI. Volume 2149 of Lecture Notes in Computer Science. (2001) 254–265

24. Uno, T., Yagiura, M.: Fast algorithms to enumerate all common intervals of two permutations. Algorithmica **26** (2000) 290–309

25. Calabrese, P.P., Chakravarty, S., Vision, T.J.: Fast identification and statistical evaluation of segmental homologies in comparative maps. ISMB (Supplement of Bioinformatics) (2003) 74–80
26. Sankoff, D., Ferretti, V., Nadeau, J.H.: Conserved segment identification. Journal of Computational Biology **4** (1997) 559–565
27. Pevzner, P., Tesler, G.: Genome rearrangements in mammalian evolution: lessons from human and mouse genomes. Genome Res **13** (2003) 37–45
28. Haas, B.J., Delcher, A.L., Wortman, J.R., Salzberg, S.L.: DAGchainer: a tool for mining segmental genome duplications and synteny. Bioinformatics **20** (2004) 3643–6
29. Vision, T.J., Brown, D.G., Tanksley, S.D.: The origins of genomic duplications in Arabidopsis. Science **290** (2000) 2114–2117
30. Bansal, A.K.: An automated comparative analysis of 17 complete microbial genomes. Bioinformatics **15** (1999) 900–908 http://www.cs.kent.edu/~arvind/orthos.html.
31. Cannon, S.B., Kozik, A., Chan, B., Michelmore, R., Young, N.D.: DiagHunter and GenoPix2D: programs for genomic comparisons, large-scale homology discovery and visualization. Genome Biol **4** (2003) R68
32. Hampson, S., McLysaght, A., Gaut, B., Baldi, P.: LineUp: statistical detection of chromosomal homology with application to plant comparative genomics. Genome Res **13** (2003) 999–1010
33. Hampson, S.E., Gaut, B.S., Baldi, P.: Statistical detection of chromosomal homology using shared-gene density alone. Bioinformatics **21** (2005) 1339–48
34. Vandepoele, K., Saeys, Y., Simillion, C., Raes, J., Peer, Y.V.D.: The automatic detection of homologous regions (ADHoRe) and its application to microcolinearity between Arabidopsis and rice. Genome Res **12** (2002) 1792–801
35. Raes, J., Vandepoele, K., Simillion, C., Saeys, Y., de Peer, Y.V.: Investigating ancient duplication events in the Arabidopsis genome. J Struct Funct Genomics **3** (2003) 117–29
36. Graur, D., Martin, W.: Reading the entrails of chickens: molecular timescales of evolution and the illusion of precision. Trends Genet **20** (2004) 80–6
37. Nei, M., Kumar, S.: Molecular Evolution and Phylogenetics. Oxford University Press (2000)
38. Zhang, L., Vision, T.J., Gaut, B.S.: Patterns of nucleotide substitution among simultaneously duplicated gene pairs in Arabidopsis thaliana. Mol Biol Evol **19** (2002) 1464–73
39. Hokamp, K.: A Bioinformatics Approach to (Intra-)Genome Comparisons. PhD thesis, University of Dublin, Trinity College (2001)
40. Bourque, G., Zdobnov, E., Bork, P., Pevzner, P., Telser, G.: Genome rearrangements in human, mouse, rat and chicken. Genome Research (2004)
41. Simillion, C., Vandepoele, K., Montagu, M.V., Zabeau, M., de Peer, Y.V.: The hidden duplication past of *Arabidopsis thaliana*. PNAS **99** (2002) 13627–32
42. O'Brien, K.P., Remm, M., Sonnhammer, E.L.L.: Inparanoid: a comprehensive database of eukaryotic orthologs. Nucleic Acids Res **33** (2005) D476–80 Version 4.0, downloaded May 2005.
43. Lynch, M., Conery, J.S.: The evolutionary fate and consequences of duplicate genes. Science **290** (2000) 1151–1155
44. Trinh, P., McLysaght, A., Sankoff, D.: Genomic features in the breakpoint regions between syntenic blocks. Bioinformatics **20 Suppl 1** (2004) I318–I325

The String Barcoding Problem is NP-Hard

Marcello Dalpasso[1], Giuseppe Lancia[2], and Romeo Rizzi[3]

[1] Dipartimento di Ingegneria dell'Informazione, University of Padova
dalpasso@dei.unipd.it
[2] Dipartimento di Matematica ed Informatica, University of Udine
lancia@dimi.uniud.it
[3] Dipartimento di Informatica e Telecomunicazioni, University of Trento
romeo.rizzi@unitn.it

Abstract. The String Barcoding (SBC) problem, introduced by Rash and Gusfield (RECOMB, 2002), consists in finding a minimum set of substrings that can be used to distinguish between all members of a set of given strings. In a computational biology context, the given strings represent a set of known viruses, while the substrings can be used as probes for an hybridization experiment via microarray. Eventually, one aims at the classification of new strings (unknown viruses) through the result of the hybridization experiment. Rash and Gusfield utilized an Integer Programming approach for the solution of SBC, but they left the computational complexity of the problem as an open question. In this paper we settle the question and prove that SBC is NP-hard.

1 Introduction

The following setting was introduced by Rash and Gusfield in [4]: Given a set V of n strings $v_1, \ldots, v_n$ (representing the genomes of n known viruses), and an extra string s (representing a virus in V, but not yet classified), we aim at recognizing s as one of the known viruses through an hybridization experiment. In the experiment, we will utilize a set P of k probes (DNA strings) and we will be able to determine which ones are contained in s (as substrings) and which are not. The result of the experiment is therefore a binary k-vector (called, in [4] a *barcode*) which can be seen as the signature of s with respect to the given probes. In order for the barcode to be able to discriminate between all the viruses, it must be true that, for each pair of viruses v_i, v_j, with $1 \leq i < j \leq n$, there exists at least one $p \in P$ which is a substring of either v_i or v_j but not of both. This amounts to saying that the barcodes of all v_i's must be distinct binary k-vectors. The cost of the hybridization experiment turns out to be proportional to k, and therefore the goal of the optimization problem, known as Minimum String Barcoding (**SBC**), is to find a feasible set P of smallest possible cardinality. Rash and Gusfield proposed an Integer Programming approach for the solution of **SBC** and also stated that a variant of the problem, in which the maximum length of each probe $p \in P$ is bounded by a constant, is NP-hard. On the other hand, they reported that "the proof for the max-length variant breaks down

A. McLysaght et al. (Eds.): RECOMB 2005 Ws on Comparative Genomics, LNBI 3678, pp. 88–96, 2005.

when you try to apply it to the unconstrained case" and they listed as the first item in the section on Future Directions: "The first item is to determine if the unconstrained **SBC** problem is NP-complete or not". In this paper we prove that **SBC** is in fact NP-complete.

The remainder of the paper is organized as follows. In Section 2 we introduce the Minimum Test Collection problem (**MTC**), a known NP-complete problem (see, e.g., Garey and Johnson [2]). In particular, we describe a special version of **MTC** which we show to be NP-complete as well, via a reduction from Set Covering. Later on, we use this special version as the starting problem to prove that **SBC** is NP-complete. In Section 3 we address the computational complexity of the problems. Subsection 3.1 introduces formally the string barcoding problems studied. In Subsection 3.2 we prove that the maximum-length version of **SBC** is NP-complete (a fact already stated in [4], although without reporting the proof, probably due to space limitations). Then, in subsection 3.3 we show how to get rid of the constraint on the substring length and we prove our main result, i.e., that **SBC** is NP-complete. The two complexity results of 3.2 and 3.3 are obtained by means of two similar reductions from **MTC**. The second time the reduction is more elaborate and starts from instances of the special case of **MTC** introduced in Section 2.

2   A Starting Problem: The Min Test Collection

In this section we introduce the Minimum Test Collection (**MTC**) problem, in its general form and in a restricted version, which we prove to be NP-complete. **MTC** and its restricted version will be used later on in the reduction to prove that **SBC** is NP-complete.

The **MTC** problem, as defined in [2], is the following problem:

MTC INSTANCE:
$D = \{d_1, \ldots, d_n\}$: a set of (ground) elements
$\mathcal{T} = \{T_1, \ldots, T_m\}$: a set of subsets of D (representing tests that may *succeed* or *fail* on the elements. A test T succeeds on d if $d \in T$ and fails on d otherwise).

MTC PROBLEM:
Find a minimum-size set $\mathcal{T}' \subseteq \mathcal{T}$ such that for any pair of elements $d, d' \in D$ there is at least one test $T \in \mathcal{T}'$ such that $|\{d, d'\} \cap T| = 1$ (i.e., the test fails on one element and succeeds on the other). A set that verifies this property is called a *testing set* of D; $\mathcal{T}'$ is a *minimum testing set* of D.

In the decision form of **MTC**, a positive integer h is also given as part of the input, and the problem requires to decide whether a testing set $\mathcal{T}' \subseteq \mathcal{T}$ with $|\mathcal{T}'| \leq h$ exists.

The **MTC** problem appears in many contexts, including one in which the elements represent a set of n diseases, and the T_i are diagnostic tests, that can verify the presence/absence of m symptoms. The goal is to minimize the number of symptoms whose presence/absence should be verified in order to correctly diagnose the disease. In [2], Garey and Johnson proved that **MTC** is NP-complete by reducing 3-dimensional Matching (**3DM**), which is NP-complete [3], to it.

We now turn to a special type of **MTC** instances, which we call *standard*. In this version of the problem, some particular tests must always be part of the problem instance.

In order to define this particular instances, assume the elements in D are ordered as $d_1, \ldots, d_n$ and let $D_j = \{d_j, \ldots, d_n\}$ for $j = 1, \ldots, n$. A set of tests T is called *suffix-closed* if $D_j \cap T \in T$ for each $T \in T$ and $j = 1, \ldots, n$. A suffix-closed set of tests T is called *standard* if $D_i \in T$ and $\{d_i\} \in T$ for each $i = 1, \ldots, n$. An instance $\langle D, T \rangle$ of **MTC** is *standard* when T is standard. In other words, a standard instance of **MTC** consists of a finite set $D = \{d_1, \ldots, d_n\}$ and a set $T = \{T_1, \ldots, T_{n(m+2)}\}$ of tests which can be partitioned as $T = T_D \cup T_I \cup T_A \cup T_E$, where

$T_D = \{T_1, \ldots, T_m\}$: a set of subsets of D;
$T_I = \{T_{m+1}, \ldots, T_{m+n}\} = \{\{d\} \mid d \in D\}$;
$T_A = \{T_{m+n+1}, \ldots, T_{m+2n}\} = \{D_j \mid 1 \le j \le n\}$;
$T_E = \{T_{m+2n+1}, \ldots, T_{n(m+2)}\} = \{T \cap D_j \mid T \in T_D, \ 2 \le j \le n\}$;

We can now prove the following result:

Theorem 1. *Minimum Test Collection (**MTC**) is NP-complete even when restricted to standard instances.*

Proof. We prove the theorem by a reduction from the Set Covering (**SC**) problem, which is defined ([1]) as follows:

SC INSTANCE:
A finite set $S = \{s_1, \ldots, s_m\}$ and a collection $\mathcal{C} = \{C_1, \ldots, C_n\} \subseteq 2^S$.
SC PROBLEM:
Find a minimum-size collection $\mathcal{C}' \subseteq \mathcal{C}$ such that every element in S belongs to at least one subset in $\mathcal{C}'$, i.e.

$$S = \bigcup_{C \in \mathcal{C}'} C \tag{1}$$

We say that any $\mathcal{C}'$ satisfying Equation 1 *covers* S, and we call such a set a *set cover* for S.

It is well known that **SC** is NP-hard. Furthermore, the version in which $\{s\} \in \mathcal{C}$ for all $s \in S$ is NP-hard as well (in fact, it is immediate to see that adding all the singletons $\{s\}$ to $\mathcal{C}$ in a **SC** leads to an **SC** instance with the exact same optimal value).

So, let $S = \{s_1, \ldots, s_m\}$ and $\mathcal{C} = \{C_1, \ldots, C_n\} \subseteq 2^S$ be an arbitrary instance of **SC**, such that $\mathcal{C}$ contains the singletons $\{s\}$ for all $s \in S$. We show how to obtain a standard instance of **MTC** representing the given instance of **SC**.

First, let $k = \lceil \log_2 m \rceil$, $K = 2^k$ and $R = \{r_1, \ldots, r_K\}$. Moreover, associate to each element $r_i \in R$ a unique binary string of length k, and associate to each $s_i \in S$ the same string of its respective element $r_i \in R$.

The set of elements D is defined as $D = R \cup S$, with a particular order: $D = \{r_1, s_1, r_2, s_2, \ldots, r_m, s_m, r_{m+1}, r_{m+2} \ldots, r_K\}$ (i.e., $D = \{d_1, \ldots, d_n\}$ with $n = m + K$). The set of tests $\mathcal{T}$ is constructed in the following way. First, for each $i = 1, \ldots, k$ we call T_i the test containing all the elements of D whose associated binary strings have the bit in position i set to 1. Then let $\mathcal{T} = \mathcal{C} \cup \{T_i \mid i = 1, \ldots, k\} \cup \{\{d\} \mid d \in D\} \cup \{D_j \mid 1 \leq j \leq n\} \cup \{T \cap D_j \mid T \in (\mathcal{C} \cup \{T_i \mid i = 1, \ldots, k\}), 2 \leq j \leq n\}$.

Now we show the following two lemmas.

Lemma 1. *If S has a set cover $\mathcal{C}' \subseteq \mathcal{C}$ of size at most h, then D has a testing set $\mathcal{T}' \subseteq \mathcal{T}$ of size at most $h + k$.*

Proof. Let $\mathcal{C}' \subseteq \mathcal{C}$ be a set cover for S of size at most h. We claim that $\mathcal{C}' \cup \{T_i \mid i = 1, \ldots, k\}$ is a testing set for D, which proves the lemma. Indeed, consider two elements s_i (or r_i) and s_j (or r_j). If $i \neq j$ then the binary strings associated to i and j differ in some position p, and hence T_p distinguishes between them. Otherwise, if $i = j$ and the two elements still differ, then we are talking about s_i and r_i; since s_i is contained in at least one set in $\mathcal{C}'$ and, moreover, no set in $\mathcal{C}'$ contains r_i, then there is some set in $\mathcal{C}'$ which distinguishes between s_i and r_i. □

Lemma 2. *If D has a testing set $\mathcal{T}' \subseteq \mathcal{T}$ of size at most h, then S has a set cover $\mathcal{C}' \subseteq \mathcal{C}$ of size at most h.*

Proof. Let $\mathcal{T}' \subseteq \mathcal{T}$ be a testing set of D of size at most h. We propose a polynomial time algorithm to produce a set $\mathcal{C}' \subseteq \mathcal{C}$ with $|\mathcal{C}'| \leq |\mathcal{T}'|$ such that $\mathcal{C}' \cup \{T_i \mid i = 1, \ldots, k\}$ is also a testing set of D. At the end, we argue that, in this case, $\mathcal{C}'$ must be a set cover of S.

Let $X = \mathcal{T}'$. Clearly, $X \cup \{T_i \mid i = 1, \ldots, k\}$ distinguishes all the elements in D, and this invariant will be maintained throughout the algorithm. If $X \subseteq \mathcal{C}$ then we just let $\mathcal{C}' = X$. Otherwise, let $T \in \mathcal{T}' \setminus \mathcal{C}$. Notice that the only pairs of elements which are not distinguished by $\mathcal{T}' \cup \{T_i \mid i = 1, \ldots, k\} \setminus \{T\}$ are of the form $\{s_i, r_i\}$; hence, T distinguishes a pair $\{s_i, r_i\}$. We now show that T can be replaced by a set of $\mathcal{C}$ who distinguishes the same couple of elements. Indeed, if T is a singleton $\{r_i\}$ for some $r_i \in R$, then it can be replaced with the respective singleton $\{s_i\}$ such that $s_i \in S$ (which is in $\mathcal{C}$ by hypothesis). Else, if T is a test D_j with $j = 2i$ and $j \leq 2K$, the ordering we have imposed among the elements of D implies that it distinguishes only the pair $s_{j/2}$ and $r_{j/2}$, so again it can be replaced with the singleton $\{s_{j/2}\}$; if T is a test $T_i \cap D_j$, a similar reasoning holds. Finally, if T is a test $C \cap D_j$ for some $C \in \mathcal{C}$, then, clearly, it can be replaced with C. Hence, by substituting every test $T \in X \setminus \mathcal{C}$ by tests in $\mathcal{C}$ as shown, we obtain that $X \subseteq \mathcal{C}$, and we let $\mathcal{C}' = X$.

We now argue simply that, since $\mathcal{C}' \cup \{T_i \mid i = 1, \ldots, k\}$ is a testing set of V, then $\mathcal{C}'$ is a set cover of S. Indeed, each pair of type $\{r_j, s_j\}$ cannot be distinguished by a set T_i, and, therefore, it must be distinguished by a test $\bar{T} \in \mathcal{C}'$. Moreover, since $r_j \notin T$ for any $T \in \mathcal{C}'$, it must be that $s_j \in \bar{T}$. Therefore, each s_j is covered, and $\mathcal{C}'$ is a set cover of S.   □

The two previous lemmas together imply that we could solve **SC** in polynomial time if and only if we could solve the corresponding **MTC** instance in polynomial time, and hence Theorem 1 is proved.   □

3   NP-Hardness of String Barcoding

3.1   The String Barcoding Problems

The following is a formal definition of the String Barcoding problem (**SBC**):

> **SBC INSTANCE:**
> An alphabet Σ (e.g., $\Sigma = \{A, C, G, T\}$) and a set $V = \{v_1, \ldots, v_n\}$ of strings over Σ (representing virus genomes).
> **SBC PROBLEM:**
> Find a minimum-size set P of strings such that for any pair of strings $v, v' \in V$ there is at least one string $p \in P$ such that p is a substring of v or v', but not both. A set that verifies this property is called a *testing set* of V; P is a *minimum testing set* of V.

Rash and Gusfield state in [4] that it is unknown whether the basic String Barcoding problem is NP-hard or not and they also state, without reporting the proof, that a variant of **SBC** called Max-length String Barcoding (**MLSBC**) is NP-hard when the underlying alphabet contains at least three elements; in this variant a constraint on the maximum length of the substrings in P is specified in input. More formally, **MLSBC** is the following problem:

> **MLSBC INSTANCE:**
> An alphabet Σ, a set $V = \{v_1, \ldots, v_n\}$ of strings over Σ and a constant k.
> **MLSBC PROBLEM:**
> Find a testing set P of V such that the length of each string $p \in P$ is less than or equal to k, and P has smallest possible cardinality among such testing sets.

3.2   An NP-Completeness Proof for MLSBC

In this section we prove that **MLSBC** is NP-hard, by considering the problem in its decision form and proving that it is NP-complete. The proof consists in reducing the Minimum Test Collection (**MTC**) problem to Max-length String Barcoding.

Theorem 2. *Max-length String Barcoding (**MLSBC**) is NP-complete.*

We reduce **MTC** to **MLSBC** in the following way.

Let $D = \{d_1, \ldots, d_n\}$ and $\mathcal{T} = \{T_1, \ldots, T_m\}$ be an instance of **MTC**. This instance can be viewed as a matrix M with n columns and m rows, in which cell $M_{ij} = 1$ if $d_j \in T_i$ (representing the fact that test T_i succeeds on element d_j), otherwise $M_{ij} = 0$. Moreover, for Ω a set of strings, we define $\bigcirc_{\omega \in \Omega} \omega$ as the string obtained as the concatenation of all the strings in Ω lined up in lexicographic order (as a matter of fact, for the purpose of our reduction to work, the strings in Ω could be concatenated in any order, but we prefer to refer to a specific order so that the instance generated through the proposed reduction is uniquely defined).

An instance of **MLSBC** is obtained in the following way. First, let $k = \lceil \log_2 m \rceil$. Then, let $\Sigma = \{A, B\}$, $\Sigma_+ = \{A, B, X\}$ (the dummy symbol X will be used as a separator, to divide the really interesting substrings, made only of As and Bs). We denote by Σ^l the set of all the strings of length l in alphabet Σ, for $1 \leq l \leq k$. Finally, uniquely encode each different element $T \in \mathcal{T}$ by a string $f_T \in \Sigma^k$ (called the *signature* of T) and let $F = \{f_T \mid T \in \mathcal{T}\}$; certainly this is possible since $|\Sigma^k| = 2^k \geq m = |\mathcal{T}|$. Now, the instance of **MLSBC** is completed by constructing the set of strings $V = \{v_d \mid d \in D\}$ such that each string $v_d \in V$ contains all the strings in Σ^{k-1} plus the signatures $f \in F$ of those tests $T \in \mathcal{T}$ that succeed on d (that is, such that $d \in T$). More formally, the codification of a disease d is the string $v_d = X^k \bigcirc_{\sigma \in \Sigma^{k-1}} (\sigma X^k) \bigcirc_{T \ni d} (f_T X^k)$.

Note that the **MLSBC** instance obtained consists of n strings $v_1, \ldots, v_n$, and the size of each such string v_i is bounded by $(2^{k-1} + m)(2k)$, which is a polynomial in m and n since $k = \lceil \log_2 m \rceil$. Therefore now it only remains to show that we can retrieve a feasible solution of one problem from a feasible solution to the other. In particular, we show that V has a testing set of size at most h if and only if D has a testing set of size at most h.

Lemma 3. *If D has a testing set $\mathcal{T}' \subseteq \mathcal{T}$ of size at most h, then V has a testing set of size at most h.*

Proof. It is easy to see that, by construction, given a testing set $\mathcal{T}' \subseteq \mathcal{T}$ for D, the set $P = \{f_T \in \Sigma^k \mid f_T \text{ is the signature of } T \in \mathcal{T}'\}$ is a testing set for V, and clearly $|P| \leq |\mathcal{T}'|$. Indeed, note that f_T is a substring of v_d if and only if $d \in T$. □

Lemma 4. *If V has a testing set of size at most h, then D has a testing set $\mathcal{T}' \subseteq \mathcal{T}$ of size at most h.*

Proof. We want to show that, given a testing set P for V, there exists a testing set $\mathcal{T}' \subseteq \mathcal{T}$ for D with $|\mathcal{T}'| \leq |P|$. In order to do this, we can assume that P is a minimal testing set, that is, $P \setminus \{p\}$ is not a testing set for all $p \in P$. We call a testing set *canonical* if all of its strings are strings in F. We claim that the minimality of P implies that P is canonical. Indeed, if P is minimal, then each $p \in P$ must distinguish some viruses. Therefore, no string in P is of the form Σ^e with $e < k$ since every such string is a substring of each $v \in V$. More generally, since the string $X^k \bigcirc_{\sigma \in \Sigma^{k-1}} (\sigma X^k)$ is contained in each $v \in V$, then P contains none of the strings X^d with $d \leq k$, nor any string obtained concatenating some string in Σ^e with some string X^d. (We also remember that $|p| \leq k$ for each $p \in P$.) This implies that every string in P belongs to Σ^*. Therefore, $|p| = k$ and so $p \in F$. Hence P is canonical. Now consider the set $T' = \{T \mid f_T \in P\}$: since P is canonical, T' is well defined and $T' \subseteq T$; moreover, $|T'| \leq |P|$ and it is easy to see that T' is a testing set.   □

Lemma 3 and Lemma 4 together conclude the reduction of **MTC** to **MLSBC** and, hence, the proof of Theorem 2.

3.3   The NP-Completeness Proof for SBC

In this section we turn our attention to **SBC** and show that it too is NP-hard. As in the case of **MLSBC**, we consider **SBC** in its decision form and prove that it is NP-complete. We show that **SBC** is NP-complete by reducing the restricted form of **MTC** to it.

Theorem 3. *String Barcoding (**SBC**) is NP-complete.*

The proof of Theorem 3 consists in a reduction of **MTC** in its restricted form to **SBC**; the reduction is the following.

Let $D = \{d_1, \ldots, d_n\}$ and $T = \{T_1, \ldots, T_{n(m+2)}\} = T_D \cup T_I \cup T_A \cup T_E$ be an instance of **MTC** of the kind described before, with $T_D = \{T_1, \ldots, T_m\}$, $T_I = \{T_{m+1}, \ldots, T_{m+n}\}$, $T_A = \{T_{m+n+1}, \ldots, T_{m+2n}\}$ and $T_E = \{T_{m+2n+1}, \ldots, T_{n(m+2)}\}$. This instance can be viewed as a matrix M with n columns and $n(m+2)$ rows, in which cell $M_{ij} = 1$ if $d_j \in T_i$ (representing the fact that test T_i succeeds on element d_j), otherwise $M_{ij} = 0$.

An instance of **SBC** is obtained, similarly to the NP-completeness proof of **MLSBC** in Section 3.2, in the following way. First, let $k = \lceil \log_2(n(m+2)) \rceil$. Then, let $\Sigma = \{A, B\}$, $\Sigma_+ = \{A, B, X\}$, and let Σ^l be the set of all the strings of length l in the alphabet Σ, for $1 \leq l \leq k$. The operator $\bigcirc$ is defined exactly in Section 3.2. Finally, uniquely encode each element $T \in T$ by a string $f_T \in \Sigma^k$ (called the *signature* of T) and let $F = \{f_T \mid T \in T\}$. Now, the instance of **SBC** is completed by constructing the set of strings $V = \{v_d \mid d \in D\}$ such that each string $v_d \in V$ contains all the strings in Σ^{k-1} plus the signatures $f \in F$ of those tests $T \in T$ that succeed on d (that is, such that $d \in T$). More formally, the codification of a disease d is the string $v_d = X^{k+j} \bigcirc_{\sigma \in \Sigma^{k-1}} (\sigma X^{k+j}) \bigcirc_{T \ni d} (f_T X^{k+j})$. Notice that the role of X is to separate the substrings, and that a different number of X characters is used in each string v in order to uniquely identify them.

The size of the constructed strings, and hence the shown transformation from the **MTC** instance to the **SBC** one, is polynomial. Therefore now it only remains to show that V has a testing set of size at most h if and only if D has a testing set of size at most h.

Lemma 5. *If D has a testing set $T' \subseteq T$ of size at most h, then V has a testing set of size at most h.*

Proof. It is easy to see that, by construction, given a testing set $\mathcal{T}' \subseteq \mathcal{T}$ for D, the set $P = \{f_T \in \Sigma^k \mid f_T \text{ is the signature of } T \in \mathcal{T}'\}$ is a testing set for V, and clearly $|P| \leq |\mathcal{T}'|$. Indeed, note that f_T is a substring of v_d if and only if $d \in T$. $\qquad\square$

Lemma 6. *If V has a testing set of size at most h, then D has a testing set $\mathcal{T}' \subseteq \mathcal{T}$ of size at most h.*

Proof. We want to show that, given a testing set P for V, there exists a testing set $\mathcal{T}' \subseteq \mathcal{T}$ for D with $|\mathcal{T}'| \leq |P|$. In order to do this, we can assume that P is a minimal testing set. Note that, by the assumption of minimality, the fact that the string $X^{k+j} \bigcirc_{\sigma \in \Sigma^{k-1}} (\sigma X^{k+j})$ is contained in each $v_j \in V$ implies that neither any of the strings of Σ^d with $1 \leq d \leq k$, nor any of the strings X^d with $1 \leq d \leq k+1$, nor any string obtained concatenating some string in Σ^d with some string X^d can be in P. More formally, we claim that P cannot contain any string of the form $X^{d_1}\{A, B\}^e X^{d_2}$ with $e < k$ and $d_1 + e + d_2 \leq k + 1$. Indeed, since P is minimal, then each $p \in P$ distinguishes some disease. Consequently the only substrings that can be contained in P have the following structure or, in turn, contain a substring with this structure:

$$\alpha = a\, X^d\, b$$

where $a \in \Sigma^{e_1}$, $b \in \Sigma^{e_2}$ and one of the following combinations of the values of d, e_1 and e_2 holds:

1. $k+1 \leq d \leq k+n$ and $e_1, e_2 > 0$; these substrings can be translated to tests from $\mathcal{T}_I$: if $d = k + j$, test T_{m+j} is taken.
2. $k+1 < d \leq k+n$ and $((e_1 = 0 \text{ and } e_2 = k) \text{ or } (e_2 = 0 \text{ and } e_1 = k))$;
   (a) if substring $f_i \in F$ (with $1 \leq i \leq m$ or $i = m + 2n$) is contained in α, these substrings can be translated to tests from $\mathcal{T}_E$: if $d = k + j$, test $T_{m+2n+(n-1)(i-1)+j-1}$ is taken;
   (b) if substring $f_i \in F$ (with $m + 1 \leq i \leq m + n$) is contained in α, these substrings can be translated to tests from $\mathcal{T}_I$: test T_i is taken;
   (c) if substring $f_i \in F$ (with $m + n + 1 < i < m + 2n$) is contained in α, these substrings can be translated to tests from $\mathcal{T}_A$: if $d = k + j$, test T_{m+n+j} is taken;
   (d) if substring $f_i \in F$ (with $m + 2n + 1 \leq i \leq n(m + 2)$) is contained in α, these substrings can be translated to tests from $\mathcal{T}_E$: if $d = k + j$, test $T_{\lfloor \frac{i-(m+2n+1)}{n-1} \rfloor (n-1)+m+2n+j-1}$ is taken.
3. $k+1 < d \leq k+n$ and $((e_1 = 0 \text{ and } 0 \leq e_2 < k) \text{ or } (e_2 = 0 \text{ and } 0 \leq e_1 < k))$; these substrings can be translated to tests from $\mathcal{T}_A$: if $d = k+j$, test T_{m+n+j} is taken.
4. $0 < d \leq k+1$ and $((e_1 = 0 \text{ and } e_2 = k) \text{ or } (e_2 = 0 \text{ and } e_1 = k))$; these substrings can be translated as follows: if substring $f_i \in F$ is contained in α, test T_i is taken.
5. $d = 0$ and $(e_1 + e_2 = k)$; as in the previous case, these substrings can be translated as follows: if substring $f_i \in F$ is contained in α, test T_i is taken.

Note that if more than one substring with this structure is contained in the considered string, the substring which is taken into account is the first one based on the ordering which is shown above.

Hence, from the strings which are contained in P we can easily obtain the elements $T \in \mathcal{T}$ which they codify, that are exactly the elements which compose the minimum testing set $\mathcal{T}'$ for D. Now consider the set $\mathcal{T}' = \{T \in \mathcal{T} \mid T$ is taken as just seen$\}$: $\mathcal{T}'$ is well defined; moreover, $|\mathcal{T}'| \leq |P|$. $\square$

Lemma 5 and Lemma 6 together conclude the reduction of **MTC** in its restricted form to **SBC** and, hence, the proof of Theorem 3.

Acknowledgments. Part of this work was supported through MIUR grants P.R.I.N. and F.I.R.B.

References

1. T. H. Cormen, C. E. Leiserson and R. L. Rivest, *Introduction to Algorithms*, MIT press, 2001.
2. M. R. Garey and D. S. Johnson, *Computers and Intractability: A Guide to the Theory of NP-Completeness*, W. H. Freeman and Co, 1979.
3. R. M. Karp. Reducibility among combinatorial problems. *Complexity and Computer Computations*, 1972.
4. S. Rash and D. Gusfield. String Barcoding: Uncovering Optimal Virus Signatures. *Proceedings of the Annual International Conference on Computational Molecular Biology (RECOMB)*, ACM Press, 2002.

A Partial Solution to the C-Value Paradox

Jeffrey M. Marcus

Department of Biology, Western Kentucky University,
1906 College Heights Boulevard #11080, Bowling Green KY 42101-1080
jeffrey.marcus@wku.edu

Abstract. In the half-century since the C-value paradox (the apparent lack of correlation between organismal genome size and morphological complexity) was described, there have been no explicit statistical comparisons between measures of genome size and organism complexity. It is reported here that there are significant positive correlations between measures of genome size and complexity with measures of non-hierarchical morphological complexity in 139 prokaryotic and eukaryotic organisms with sequenced genomes. These correlations are robust to correction for phylogenetic history by independent contrasts, and are largely unaffected by the choice of data set for phylogenetic reconstruction. These results suggest that the C-value paradox may be more apparent than real, at least for organisms with relatively small genomes like those considered here. A complete resolution of the C-value paradox will require the consideration and inclusion of organisms with large genomes into analyses like those presented here.

1 Introduction

In the years following the discovery that DNA was the hereditary material [1], and even before the structure of DNA was fully understood [2], investigators measured the amount of haploid DNA (or C-value) in the cells of various organisms, hoping that this quantity might provide insights into the nature of genes [3]. They found no consistent relationship between the amount of DNA in the cells of an organism and the perceived complexity of that organism, and this lack of correspondence became known as the C-value paradox [4].

The C-value paradox has become one of the enduring mysteries of genetics, and generations of researchers have repeatedly referred to the lack of correspondence between genome size and organismal complexity [3, 5-9]. In spite of all of the attention devoted to the C-value paradox over more than five decades, there has yet to be an explicit statistical correlation analysis between measures of genome size and measures of organismal morphological complexity. Organismal complexity has been difficult to examine rigorously because of the inherent difficulties in measuring the complexity of organisms. Rather than trying to measure morphological complexity, most researchers studying the C-value paradox referred explicitly or implicitly to a complexity scale with bacteria at the bottom; protists, fungi, plants, and invertebrates in the middle; and vertebrates (particularly humans) at the top (Figure 1). This scale, called the Great Chain of Being, can be traced back to Aristotle and exerts a pervasive

A. McLysaght et al. (Eds.): RECOMB 2005 Ws on Comparative Genomics, LNBI 3678, pp 97-105, 2005
© Springer-Verlag Berlin Heidelberg 2005

influence over popular and scientific perceptions of complexity [10]. This scale is not quantitative and incorporates untested assumptions about the relative complexity and monophyly of taxonomic groups.

Instead, researchers have focused on studying the statistically significant relationships between genome size and a variety of other quantitative traits such as cell volume, nuclear volume, length of cell cycle, development time, and ability to regenerate after injury [5, 11-15], which had long been included as part of the C-value paradox [3, 4], but which recently have collectively been redefined as a distinct phenomenon known as the C-value enigma [6]. There have been two general categories of hypotheses concerning the cause of the C-value enigma and of genome size variation [5, 6]. Explanations in the first category suggest that the bulk of the DNA has an adaptive significance independent of its protein-coding function. The amount of DNA may affect features such as nuclear size and structure or rates of cell division and development, suggesting that changes in genome size may be adaptive [6, 12, 13]. The second category of explanation suggests that the accumulation of DNA is largely nonadaptive, and instead represents the proliferation of autonomously replicating elements that continue to accumulate until the cost to the organism becomes significant [16, 17]. Gregory [6, 14, 15] has recently summarized the available data and argues that variation in genome size (and by implication variation in amounts of genomic heterochromatin) is predominantly due to direct selection on the amount of bulk DNA via its causal effects on cell volume and other cellular and organismal parameters.

The work described in this paper is explicitly not an examination of the C-value enigma, which is relatively well studied. This study attempts to address a very different question, the C-value paradox *sensu stricto* or the relationship between genome size and organismal morphological complexity, which is virtually unstudied. There are several developments that have increased the tractability of this type of investigation. One of the most important of these is the availability of many organisms with sequenced genomes, providing us with reliable estimates of both genome size and number of open reading frames (an estimate of gene number) [18]. A second advance has been the development of measures of non-hierarchical morphological complexity [19]. The number of cell types produced by an organism is among the most commonly used indices of non-hierarchical morphological complexity, and there are cell type counts available for a wide variety of organisms [20-24]. A final advance has been the development of comparative techniques such as phylogenetically independent contrast analysis [25-27]. Phylogenetically independent contrasts allows the study of correlations among traits between different species of organisms, even though the organisms vary in their degree of relatedness and are therefore not independently and identically distributed. It does this by using an explicit phylogeny to create a series of contrasts between pairs of sister taxa which, by definition, are the same age, so the time elapsed and the accumulated phylogenetic distance between the sister taxa is factored out of the analysis. The resulting contrasts are independently and identically distributed, and therefore suitable for correlation analysis. The novel approach presented here builds on these developments, using measures of genome size and complexity from sequenced genomes, numbers of cell types and numbers of subcellular parts as measures of morphological complexity, and phylogenetically independent contrast analysis to provide the first explicitly statistical analysis of the C-value paradox. The results of independent contrast analysis suggest that the C-value and measures of morphological complexity are significantly positively correlated.

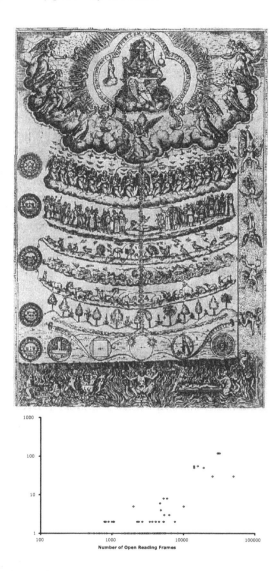

Fig. 1. Representations of the relative complexity of organisms. Above: Pictorial representation of the Great Chain of Being as depicted in Valades [28] (after Fletcher [29]). Below: the correspondence between number of open reading frames (an estimate of gene number) and the number of cell types produced for the first 139 organisms with sequenced genomes

2 Materials and Methods

While genome size has been the preferred metric for comparison with complexity, at least initially it was intended to be a proxy for gene number [3], which was difficult to estimate accurately until whole genome sequencing became possible. Genome size and number of open reading frames (an estimate of gene number) for the first 139 completely sequenced genomes (including 121 prokaryotes and 18 eukaryotes) were obtained from two genome databases [30, 31]. The number of cell types and for prokaryotes, the number of types of cell parts, produced by each organism was obtained from the literature (On-Line Supplementary Table 1). Cell types were considered distinct if intermediate morphologies were rare or absent. Counts of types of cell parts were determined from descriptions of prokaryote ultrastructure that were included in species descriptions. The number of cell types and the number of types of cell parts represent non-hierarchical indices of complexity—organisms with more cell types or cell parts are considered to be more complex than organisms with fewer cell types or cell parts [19]. To control for evolutionary relatedness that might confound correlations between these measures, I used independent contrast analysis with phylogenetic trees generated from the small subunit of ribosomal RNA [32], using sequences available for each taxon from NCBI [33]. Sequences were aligned in CLUSTALX [34] and phylogenetic trees were generated by Neighbor-Joining, Parsimony, and Maximum Likelihood methods as implemented in PAUP* [35] with Eukaryote 18s rRNA sequences used as the outgroup.

In addition, a data set for 45 taxa (4 eukaryotes and 41 prokaryotes), compiled and aligned by Brown et al. [36], and consisting of amino acid sequences for 23 conserved genes was kindly provided by James R. Brown. Open reading frame counts for two of the species included in Brown et al. [36], *Porphyromonas gingivialis* and *Actinobacillus actinomycetemosomonas* were not available when these analyses were conducted, so these species were excluded in my analysis (leaving 43 taxa: 4 eukaryotes and 39 prokaryotes). The Brown et al. (2001) data set was analyzed using Neighbor-Joining and Parsimony techniques. Jukes-Cantor branch lengths were applied to all trees, branch lengths of 0 were converted to 0.000001, and all branch lengths were transformed to the square root of the Jukes-Cantor distance to standardize them for analysis by contrasts [27].

3 Results

The evolutionary trees produced by the phylogenetic analyses are not shown because they largely replicate the results of Nelson et al. [37] and Brown et al. [36]. These analyses had to be repeated because independent contrast analysis requires that the species included in the phylogeny and the species included in the continuous character data sets must be completely congruent. Independent contrast analyses using trees produced by different tree-building algorithms from the same data set produced highly similar correlation coefficients, while analyses using trees derived from different data sets had larger differences in correlation coefficients. However, the significance of independent contrast correlations was generally robust to changes in tree topology, suggesting that phylogenetic uncertainty due to tree differences is unimportant in interpreting these correlations [38].

3.1 Small Subunit rRNA Phylogeny Independent Contrast Analysis

First, the small subunit rRNA data set will be considered. After independent contrast analysis, the correlation between number of cell types and genome size was significant or nearly significant depending on the tree used (Table 1), and the correlation between the number of cell types and the number of open reading frames was highly significant (Figure 1). As eukaryotes, with their larger genome size and greater number of cell types, might unduly influence these results, so the eukaryotes were pruned from the trees and the independent contrast analysis was repeated. A significant correlation was detected between the number of cell types and genome size, as was the number of cell types and number of open reading frames. To answer the concern that prokaryote cell diversity might be better expressed in terms of numbers of cell parts (organelle-like structures: prokaryotes by definition do not have true organelles), rather than numbers of cell types, the number of types of cell parts for each prokaryote was also collected. This yielded a significant correlation between the number of types of cell parts and genome size and between the number of types of cell parts and the number of open reading frames.

Table 1. Independent contrast analyses for the small subunit rRNA phylogeny

Indpendent Contrast	N	r	p
With Eukaryotes			
Genome size vs. Number of Cell Types	139	0.155-0.186	0.029-0.068
ORFs vs. Number of Cell Types	139	0.616-0.641	<0.0001
Without Eukaryotes			
Genome size vs. Number of Cell Types	121	0.225-0.228	0.011-0.013
ORFs vs. Number of Cell Types	121	0.192-0.197	0.030-0.034
Genome size vs. Number of Cell Parts	121	0.278-0.283	0.002
ORFs vs. Number of Cell Parts	121	0.276-0.277	0.002

3.2 Conserved Gene Amino Acid Phylogeny Independent Contrast Analysis

Substantially similar relationships were found using alternative phylogenetic trees derived from conserved protein sequences for 43 species [36], suggesting that these correlations are not an artifact of trees derived from small subunit rRNA sequences. The independent contrast analysis using the Brown et al. (2001) data set showed a significant positive correlation between number of cell types and genome size and between the number of cell types and the number of open reading frames (Table 2). Pruning eukaryotes from the trees and repeating the analysis yielded significant correlations between number of cell types and genome size and between the number of cell types and the number of open reading frames. Continuing to restrict the analysis to prokaryotes and considering the number of types of cell parts gave significant or nearly significant correlations between this quantity and both genome size and the number of open reading frames size.

Table 2. Independent contrast analyses for the conserved gene amino acid phylogeny

Indpendent Contrast	N	r	p
With Eukaryotes			
Genome size vs. Number of Cell Types	43	0.785-0.800	<0.0001
ORFs vs. Number of Cell Types	43	0.899-0.908	<0.0001
Without Eukaryotes			
Genome size vs. Number of Cell Types	39	0.438-0.462	0.004-0.005
ORFs vs. Number of Cell Types	39	0.432-0.459	0.003-0.006
Genome size vs. Number of Cell Parts	39	0.308-0.320	0.047-0.056
ORFs vs. Number of Cell Parts	39	0.345-0.360	0.024-0.031

4 Discussion

For all of the data sets examined here, there are significant positive correlations between genome size or numbers of open reading frames and numbers of cell types and numbers of types of cell parts. These results suggest that the greatest irony about the C-value paradox may very well be that there is no paradox at all and that genome complexity and morphological complexity actually do significantly positively correlate with one another, at least for the organisms with sequenced genomes in this data set. This is not the first time a correspondence between genome size and morphological complexity has been suggested [16, 39, 40], but this is the first time the correspondence is supported by an analysis of independent contrasts that reveals a statistically significant positive correlation.   This suggests that organismal morphological complexity may follow some of the same scaling laws that have already been observed in other combinatorial systems [41].

While these results differ from those of most previous studies of the C-value paradox, previous methods for measuring these quantities (such as haploid DNA content, chromosome number, or placement on the scale of the Great Chain of Being [10]) may have been inadequate to detect these correlations. The development of whole genome sequencing and annotation [30, 31] and the creation of new metrics for measuring complexity [19] have permitted this finer-scale understanding of the relationship between morphological complexity and genomic complexity. For those interested in the relationship between genotype and the generation of morphological complexity [42], the detected correlations between numbers of open reading frames and numbers of cell types or types of cell parts suggest that the number of genes present in an organism may have a greater role in permitting, generating, or maintaining morphological complexity than previously anticipated.

A note of caution is warranted in interpreting these results because the selection of genomes to be sequenced has been influenced by genome size, because larger genomes are more costly to sequence. As a result, the tendency has been to select, particularly among eukaryotes, morphologically complex organisms with the smallest possible genome sizes for sequencing. This could predispose data sets containing eukaryotes to reveal positive correlations between genome size and morphological complexity because of issues of taxon sampling. However, the selection of prokaryotes for sequencing, because of their universally much smaller genome sizes, is largely free from this bias, so the analyses of the prokaryote-only data sets included here are probably revealing real positive correlations between measures of genome size and complexity and measures of morphological complexity.

Complete resolution of the C-value paradox will require the consideration of eukaryotic organisms with large genomes and significant amounts of heterochromatin so that a determination can be made concerning whether the relationships reported here also hold at larger genome sizes, something that may not be possible until several organisms with large genomes have been sequenced.

Acknowledgements

The author thanks D. McShea and R. Fehon for the public discussion that inspired this paper, B. Nicklas for his inspirational lectures on genome size, R. Vilgalys for insights into microbial diversity and ultrastructure, J. Brown for the aligned amino acid data set, J. Shapiro and T. Carter for their encouragement, and J. Seiff and B. Marcus for their help and patience. D. McShea, K. Doerner, D. Emlen, T. Evans, A. Harper, K. Hertweck, T. Hughes, T. Powell, B. Polen, and two anonymous reviewers provided helpful comments on the manuscript. Support for this research is from the National Science Foundation and the Commonwealth of Kentucky through an EPSCoR award (EPS-0132295), from the National Institutes of Health and the National Center for Research Resources Grant P20 RR16481, from a scholarship to the Complex Systems Summer School at the Santa Fe Institute, and from a Junior Faculty Scholarship from Western Kentucky University.

References

1. Avery, O.T., C.M. MacLeod, and M. McCarty, Studies on the Chemical Nature of the Substance Inducing Transformation of Pneumococcal Types: Induction of Transformation by a Deoxyribonucleic Acid Fraction Isolated from Pneumococcus Type III. J. Exp. Med., 1944. **79**(1): p. 137-158.
2. Watson, J.D. and F.H.C. Crick, A structure for Deoxyribose Nucleic Acid. Nature, 1953. **171**: p. 737-738.
3. Mirsky, A.E. and H. Ris, The deoxyribonucleic acid content of animal cells and its evolutionary significance. J. gen. Physiol., 1951. **34**: p. 451-462.
4. Thomas, C.A., The genetic organization of chromosomes. Annu. Rev. Genet., 1971. **5**: p. 237-256.
5. Cavalier-Smith, T., ed. The evolution of genome size. 1985, John Wiley: New York.
6. Gregory, T.R., Coincidence, coevolution, or causation? DNA content, cell size, and the C-value enigma. Biol. Rev., 2001. **76**: p. 65-101.
7. Pagel, M. and R.A. Johnstone, Variation across species in the size of the nuclear genome supports the junk-DNA explanation for the C-value paradox. Proc. R. Soc. Lond., 1992. **249**: p. 119-124.

8. Goin, O.B., C.J. Goin, and K. Bachmann, DNA and amphibian life history. Copeia, 1968. **1968**: p. 532-540.
9. Ohno, S., Evolution by gene duplication. 1970, New York: Springer-Verlag.
10. Lovejoy, A.O., The Great Chain of Being. 1936, Cambridge, MA: Harvard University Press. 376.
11. Cavalier-Smith, T., Nuclear volume control by nucleoskeletal DNA, selection for cell volume and cell growth rate, and the solution of the DNA C-value paradox. J. Cell Sci., 1978. **43**: p. 247-278.
12. Cavalier-Smith, T., r- and K-tactics in the evolution of protist developmental systems: cell and genome size, phenotype diversifying selection, and cell cycle patterns. Biosystems, 1980. **12**: p. 43-59.
13. Sessions, S.K. and A. Larson, Developmental correlates of genome size in plethodontid salamanders and their implications for genome evolution. Evolution, 1987. **41**: p. 1239-1251.
14. Gregory, T.R., Genome size and developmental complexity. Genetica, 2002. **115**: p. 131-146.
15. Gregory, T.R., Macroevolution, hierarchy theory, and the C-value enigma. Paleobiology, 2004. **30**(2): p. 179-202.
16. Doolittle, W.F. and C. Sapienza, Selfish genes, the phenotype paradigm and genome evolution. Nature, 1980. **284**: p. 601-603.
17. Orgel, L.E. and F.H.C. Crick, Selfish DNA: the ultimate parasite. Nature, 1980. **284**: p. 604-607.
18. Nelson, K.E., Paulsen, I.T., Heidelberg, J.F., and Fraser, C.M., Status of genome projects for nonpathogenic bacteria and archaea. Nature Biotechnology, 2000. **18**: p. 1049-1054.
19. McShea, D.W., Functional complexity in organisms: Parts as proxies. Biol. Philos, 2000. **15**(5): p. 641-668.
20. Sneath, P.H.A., Comparative biochemical genetics in bacterial taxonomy, in Taxonomic Biochemistry and Serology, C.A. Leone, Editor. 1964, Ronald Press: New York. p. 565-583.
21. Valentine, J.W., A.G. Collins, and C. Porter Meyer, Morphological complexity increase in metazoans. Paleobiology, 1994. **20**(2): p. 131-142.
22. Carroll, S.B., Chance and necessity: the evolution of morphological complexity and diversity. Nature, 2001. **409**(6823): p. 1102-1109.
23. Bell, G. and A.O. Mooers, Size and complexity among multicellular organisms. Biol. J. Linn. Soc., 1997. **60**: p. 345-363.
24. Bonner, J.T., The evolution of complexity by means of natural selection. 1988, Princeton, NJ: Princeton University Press. 260.
25. Harvey, P.H. and M.D. Pagel, The comparative method in evolutionary biology. 1991, Oxford: Oxford University Press.
26. Felsenstein, J., Phylogenies and the comparative method. Am. Nat., 1985. **125**: p. 1-15.
27. Garland, T., Jr., P.H. Harvey, and I.R. Ives, Procedures for the analysis of comparative data using phylogenetically independent contrasts. Syst. Biol., 1992. **41**: p. 18-32.
28. Valades, D., Rhetorica Christiana. 1579: Pervsiae, apud Petrumiacobum Petrutium. 10.
29. Fletcher, A., Gender, Sex, and Subordination in England 1500-1800. 1995, New Haven: Yale University Press. 442.
30. CBS Genome Atlas Database. 2003, Center for Biological Sequence Analysis, http://www.cbs.dtu.dk/services/GenomeAtlas/: Lyngby, Denmark.
31. GOLD Genomes OnLine DataBase. 2003, Integrated Genomics, http://igweb.integratedgenomics.com/GOLD/: Chicago, IL.

32. Martins, E.P., COMPARE, version 4.4. Computer programs for the statistical analysis of comparative data. 2001, Department of Biology, Indiana University, Bloomington IN.
33. National Center for Biotechnology Information. 2003, National Library of Medicine, http://www.ncbi.nlm.nih.gov/: Washington, D.C.
34. Jeanmougin, F., et al., Multiple sequence alignment with Clustal X. Trends Biochem. Sci., 1998. **23**: p. 403-405.
35. Swofford, D.L., PAUP*, Phylogenetic analysis using parsimony (*and other methods). 1998, Sinauer Associates: Sunderland, Massachusetts.
36. Brown, J.R., et al., Universal trees based on large combined protein sequence data sets. Nat. Genet., 2001. **28**: p. 281-285.
37. Nelson, K.E., et al., Status of genome projects for nonpathogenic bacteria and archaea. Nature Biotechnology, 2000. **18**(10): p. 1049-1054.
38. Marcus, J.M. and A.R. McCune, Ontogeny and phylogeny in the northern swordtail clade of Xiphophorus. Syst. Biol., 1999. **48**(3): p. 491-522.
39. Rees, H. and R.N. Jones, The origin of the wide species variation in nuclear DNA content. Int. Rev. Cytol., 1972. **32**: p. 53-92.
40. Sparrow, A.H., H.J. Price, and A.G. Underbrink, A survey of DNA content per cell and per chromosome of prokaryotic and eukaryotic organisms: some evolutionary considerations. Brookhaven Symp. Biol., 1972. **23**: p. 451-494.
41. Changizi, M.A., Universal Scaling Laws for Hierarchical Complexity in Languages, Organisms, Behaviors and other Combinatorial Systems. J. Theor. Biol., 2001. **211**: p. 277-295.
42. Hedges, S.B., et al., A molecular timescale of eukaryote evolution and the rise of complex multicellular life. BMC Evol. Biol., 2004. **4**: p. 2 doi:10.1186/1471-2148-4-2.

Individual Gene Cluster Statistics in Noisy Maps

Narayanan Raghupathy[1,*] and Dannie Durand[2]

[1] Department of Biological Sciences, Carnegie Mellon University,
Pittsburgh, PA 15213, USA
rnarayan@cmu.edu

[2] Departments of Biological Sciences and Computer Science,
Carnegie Mellon University, Pittsburgh, PA 15213, USA
durand@cmu.edu

Abstract. Identification of homologous chromosomal regions is important for understanding evolutionary processes that shape genome evolution, such as genome rearrangements and large scale duplication events. If these chromosomal regions have diverged significantly, statistical tests to determine whether observed similarities in gene content are due to history or chance are imperative. Currently available methods are typically designed for genomic data and are appropriate for whole genome analyses. Statistical methods for estimating significance when a single pair of regions is under consideration are needed. We present a new statistical method, based on generating functions, for estimating the significance of orthologous gene clusters under the null hypothesis of random gene order. Our statistics is suitable for noisy comparative maps, in which a one-to-one homology mapping cannot be established. It is also designed for testing the significance of an individual gene cluster in isolation, in situations where whole genome data is not available. We implement our statistics in Mathematica and demonstrate its utility by applying it to the MHC homologous regions in human and fly.

1 Introduction

Identification of pairs of homologous chromosomal regions is an important step in solving a broad range of evolutionary and functional problems that arise in comparative mapping and genomics. Closely related homologous regions will be characterized by conserved gene order and content, and may have substantial similarity in non-coding regions as well. However, in more distantly related regions, significant sequence similarity will typically only be observable in coding regions. In this case, genes are frequently treated as markers and putative homologous regions are identified by searching for *gene clusters*, regions that share similar gene content but where neither content nor order are preserved. Statistical tests to distinguish significant clusters from chance similarities in gene organization become essential as gene content and order diverge.

Conserved regions in whole genome comparisons are the basis of comparative map construction, studies of genome rearrangements [1–3] and gene order

---

* Contact author.

A. McLysaght et al. (Eds.): RECOMB 2005 Ws on Comparative Genomics, LNBI 3678, pp. 106–120, 2005.
© Springer-Verlag Berlin Heidelberg 2005

conservation [4–6], alternative approaches to phylogeny reconstruction [5, 7–11] and operon prediction in prokaryotes [12,13]. Genome self-comparison is used to test hypotheses of whole genome duplication [14,15]. Studies such as these consider gene clusters in a genomic context, focusing on large scale evolutionary processes and chromosomal organization.

In addition, many evolutionary and functional studies are based on studies of a single linkage groups [16–28]. Some studies examine the evolutionary history of a particular conserved region and the selective forces that hold it together. Others seek to exploit local similarities in gene organization for functional inference, gene annotation or to disambiguate orthology identification. For organisms whose lifestyle (e.g., lamprey) or longevity (e.g., fig) precludes construction of a genetic map, further research depends on identification of a homologous region in a species that is more suited to genetic manipulation or metabolic studies.

Analyses of such *individual clusters* often cannot take broader genomic context into consideration. For example, many researchers in fields such as ecology and organismal, behavioral and evolutionary biology work on species which have not been sequenced and are unlikely to be sequenced in the foreseeable future. In such cases, the amount of information about a region of interest, is limited by the laboratory's sequencing budget and available sequences in public data bases.

Our goal is to develop methods for estimating the statistical significance of individual gene clusters that can be carried out with knowledge of a local region plus aggregate properties such as an estimate of total gene number.

1.1 Related Work

While statistical models for testing cluster significance are beginning to appear in the literature [21, 29–33], none are currently suitable for testing the significance of individual clusters. The lack of genomic context imposes a number of constraints on the statistical approach. Monte Carlo methods typically involve randomization of the entire genome, which is possible only with a complete genomic data set. When this information is not available, analytical tests that are parameterized by aggregate properties (e.g., the size of the genome, the size and number of gene families, etc.) are required.

The statistical tests for individual clusters must be based on an appropriate cluster definition. The intuitive notion that gene clusters share similar but not conserved gene content has been translated into a number of different formal models for finding and testing clusters [6, 29, 33–45], yet most of these are not suitable for individual clusters. Cluster statistics depend on the size of the search space. A number of statistical tests have been developed for a *reference region* model [31,46,21,33,47], in which an investigator is interested in a particular genomic region and searches the entire genome for additional regions containing the same genes. If the total number of genes in the genome is n and we are interested in m reference genes ($m << n$), then there are $n - m + 1$ regions to be considered. Thus in the reference region scenario $O(n)$ pairs of regions must be compared. However, when the investigator selects one or more pairs of homologous anchor genes and searches their genomic neighborhoods for additional homologs, the search space is $O(1)$. For such studies, the $O(n)$ reference region approach will underestimate the significance of the cluster.

Furthermore, the cluster definition must not require whole genome context to make sense. Many studies are based on the *max-gap* cluster, a maximal set of homologous pairs where the distance between adjacent homologs on the same chromosome is no greater than a pre-specified parameter, g. The nature of the max-gap definition creates a "look-ahead" problem [36,46,45], such that maximal max-gap clusters cannot be found using local, greedy heuristics. Although a local search may suggest that a particular region does not contain a max-gap cluster, a whole genome search is required to verify that a cluster meeting the max-gap criterion does not exist. Thus, while statistical tests for max-gap clusters based on the reference region model have been developed [46], these are not appropriate for individual clusters found by local search.

Finally, most current statistical tests do not take gene families into account, yet the significance of a cluster decreases as gene family size grows, because a given gene in one genome can be homologous to more than one gene in the other. As the number of possible matches increases, so do chance occurrences of gene clusters. As a result, tests that do not take gene family size into account risk overestimation of cluster significance.

1.2   Results

We propose a statistical test for individual clusters, under the null hypothesis of random gene order. These may be used without complete genomic context, are suitable for individual clusters found by local search, incorporate gene family size and are computationally tractable. In previous work [31], we proposed a test for individual clusters based on a window sampling model. Given two genomes with gene families, our measure of significance was the probability of observing a conserved set of linked genes in close proximity on both genomes. However, the treatment presented was mainly of theoretical interest since it did not lend itself to a computationally tractable implementation.

In the current paper, we recast this model in terms of generating functions, allowing us to obtain a general expression for our test statistic under the assumption of arbitrary gene family sizes. This statistic requires only the size of the conserved region, number of homologous genes in the linkage group and estimates of the distribution of gene family sizes and of the total number of genes in the genome. No information about the spatial organization of the genome outside the conserved region is needed. Under the additional assumption of fixed gene family sizes, we use the generating function model to obtain closed form expressions approximating cluster probabilities that can be calculated efficiently using Mathematica.

We describe our model and give a formal statement of the problem in Section 2. In Section 3, we derive the probability of observing an individual cluster under the null hypothesis of random gene order.

In Section 4, we demonstrate how our model may be applied, using the heavily studied conserved homologous regions associated with Major Histocompatibility Complex (MHC) in human and fly.

2   Model

We develop tests for individual clusters based on the probability of observing a cluster in a genome with uniform random gene order (a "random genome"). A genome, $G_i = (1, \ldots, n_i)$ is modeled as an ordered set of n_i genes, ignoring chromosome break and physical distances between genes. We assume that genes do not overlap.

2.1   The 1-1 Model

We begin with a simple model of two genomes, G_1 and G_2, with identical gene content and a one-to-one mapping between genes in G_1 and genes in G_2. That is, every gene in G_1 has exactly one homolog in G_2 and vice versa. We define an orthologous cluster as a pair of windows, W_1 and W_2, of length r_1 and r_2 selected from genomes G_1 and G_2, respectively, that share m homologous gene pairs. Figure 1 shows a cluster of three genes in a window of size five.

$$\cdots \bullet \bullet \bullet \, (\, v \; w \; \bullet \bullet \, u \,) \, \bullet \bullet \bullet \bullet \bullet \bullet \bullet \bullet \cdots$$
$$\cdots \bullet \bullet \bullet \bullet \bullet \bullet \bullet \, (\, u \; \bullet \; w \; v \, \bullet) \, \bullet \bullet \bullet \bullet \cdots$$

Fig. 1. A cluster with $r = 5$, $m = 3$ in the *1-1* model. Genes without homologs in this region shown as dots.

In this simple model, the probability that a pair of windows, of length r_1 and r_2, have exactly m genes in common is simply the probability that m of the r_1 genes in W_1 also appear in W_2 and can be calculated using a hypergeometric distribution:

$$p_{1-1}(m) = \frac{\binom{r_1}{m}\binom{n_2-r_1}{r_2-m}}{\binom{n_2}{r_2}}$$

The probability that the windows share *at least* m genes is then $\sum_{i=m}^{r} p_{1-1}(i)$. The 1-1 model requires a perfect, unambiguous homology mapping between G_1 and G_2. This may be possible after a recent speciation or polyploidization event. In general, however, because of variations in mutation rates, convergent evolution, non-homologous gene displacement and multi-domain proteins generated by exon shuffling, it is not possible to identify a unique match.

In this case, a many-to-many model is required. Genes are partitioned into families, such that any gene in a given family in G_1 can match any gene in the same family in G_2. The probability of finding a cluster by chance increases with family size. Consider, for example, the simple scenario where just one of the genes in W_1 matches f genes in G_2. The probability of finding m matches to the genes in a fixed size in G_2 increases since there are f possible matches for this one gene. However, it is surprisingly difficult to obtain a straightforward closed formula expressing this probability, even for this simple scenario. Therefore, accurate statistics require a model of gene family size. However, this raises the challenge that once gene families are incorporated in the model, it is no longer easy to determine the expected number of matches in a window of size r.

2.2   General Gene Family Model

The problem of identifying true homologs has been much debated and numerous solutions have been proposed [33, 43, 48–50]. The first step is typically sequence comparison. A variety of approaches are applied to rule out false positives or negatives due to weak sequence similarity and/or matches based on homologous domains in otherwise unrelated sequences. These include bi-directional best hits, imposing a minimum alignment length requirement and phylogeny reconstruction. Despite these efforts, homology frequently remains unresolved. Furthermore, gene duplications that occur after the speciation separating G_1 and G_2, result in situations where a gene in one genome has two or more legitimate orthologs in the other.

We therefore extend our model to include gene families. A gene family is a set of *homologous genes*; that is, genes that share a common ancestor, through either duplication (paralogs) or speciation (orthologs). Gene family membership in our model does not depend on inherent functional or structural properties of the family but rather on what type of information the user brings to bear on identification of homology relationships. We define a gene family to be the set of *indistinguishable* homologous genes; i.e., homologous genes, where subfamily classification cannot be further disambiguated.

This is illustrated by the tyrosine kinases, a large multi-domain family of eukaryotic signaling proteins with 90 members in human [51]. While sequence similarity in the kinase domain shows that all tyrosine kinases are related, the domain composition of these sequences varies greatly, so that domain architecture can be used to disambiguate orthology. For instance, the Insulin Receptor (IR), in addition to the kinase domain, has two Furin-like domains, a Leucine-rich domain and two fibronectin domains, a domain architecture shared only with two other human genes: the Insulin Growth Factor 1 Receptor and the Insulin-Receptor Related Receptor. Thus, while an analysis based on sequence comparison alone might map mouse IR to almost 100 kinases in human, an analysis based on domain architecture would associate mouse IR with only three human homologs.

We will assume that the set of genes in genomes G_1 and G_2 can be partitioned into non-intersecting gene families. Let $f_{ij} \subset G_i$ denote the members of the jth gene family in genome i. Then, the jth gene family, $f_j = f_{1j} \cup f_{2j}$, is a set of genes such that each gene in f_j is homologous to all other genes in f_j and only those genes. There are $\phi_{ij} = |f_{ij}|$ genes in the jth family in genome G_i. Let $\mathcal{F} = \{f_j\}$ be the set of all gene families in both genomes. In the gene family model, we define an orthologous gene cluster to be a pair of windows of length r_1 and r_2, drawn from G_1 and G_2, respectively, that have m *gene families* in common.

3   Cluster Statistics

We develop a test for individual clusters based on the probability of observing a cluster in two genomes with uniform random gene order (a "random genome"). In calculating cluster probabilities for the general case, we will need to count the number of ways that a window of a particular size can be filled with a given set of gene families in several contexts. We therefore derive a general solution to this problem using generating functions, a powerful combinatorial approach which can be used to determine a sequence of interest from the coefficients of a power series(see, for example, [52]). Here the sequence of interest is the number of ways filling the window. It is this formalism that allows us to compute cluster probabilities efficiently.

3.1   Window Packings

Define $\mathcal{T}$ to be a set of λ gene families of arbitrary size $\phi_1 \ldots \phi_\lambda$. Given the sample space of all sets of w genes sampled from a genome of size n, we wish to enumerate those that contain at least one gene from each family in $\mathcal{T}$. Since we do not take into account the order of genes in a window, this enumeration is equivalent to finding all window packings. The generating function formulation allows us to determine the number of such window packings, denoted by $\mathcal{N}(w, \lambda, \mathcal{T})$.

We represent contribution of the jth family in $\mathcal{T}$ by the generating function

$$\alpha_j(t) = \binom{\phi_j}{1} t + \binom{\phi_j}{2} t^2 + \ldots + \binom{\phi_j}{\phi_j} t^{\phi_j}. \tag{1}$$

The coefficient of t^i in $\alpha_j(t)$, denoted by $[t^i]\alpha_j(t)$, represents the number of ways of choosing i genes from j^{th} family. The contributions of all λ families to the window can then be derived from the product of their generating functions:

$$\alpha(t) = \prod_{j=1}^{\lambda} \left[\binom{\phi_j}{1} t + \binom{\phi_j}{2} t^2 + \ldots + \binom{\phi_j}{\phi_j} t^{\phi_j} \right]. \tag{2}$$

The coefficient $[t^w]\alpha(t)$ gives the number of ways of filling w slots with genes from the λ families, which is just $\mathcal{N}(w, \lambda, \mathcal{T})$. Note that the t^w term in $\alpha(t)$ will be a sum of products of the form $\beta_1 t^{x_1} \cdot \beta_2 t^{x_2} \cdots \beta_\lambda t^{x_\lambda} = (\prod_j \beta_j) t^w$, where the exponents of the dummy variable, t, sum to w. By inspecting Equation (2), we see that since β_j is the coefficient of t^{x_j}, it must be of the form $\beta_j = \binom{\phi_j}{x_j}$. The term $[t^w]\alpha(t)$ corresponds to packings containing x_1 genes from the first family, x_2 genes from the second family and so forth, where β_j corresponds to the number of ways of choosing x_j genes from the jth gene family. Summing over all packings, we obtain

$$\mathcal{N}(w, \lambda, \mathcal{T}) = \sum_{(x_1, \cdots x_\lambda)} \binom{\phi_1}{x_1} \binom{\phi_2}{x_2} \cdots \binom{\phi_\lambda}{x_\lambda}, \tag{3}$$

where the sum is over the set of all λ-tuples $(x_1, \ldots, x_\lambda)$ such that

$$\sum_{j=1}^{\lambda} x_j = w, \tag{4}$$

and $1 \leq x_j \leq \phi_j, \forall j$.

Let us illustrate the window packing problem with a simple example. Suppose we wish to find the number of ways a window of size $w = 7$ can be packed with four gene families ($\lambda = 4$), such that the window has at least one gene from each gene family. Let the gene family sizes of T be $\phi_1 = 1$, $\phi_2 = 2$, $\phi_3 = 3$ and $\phi_4 = 4$ and the 4-tuple (x_1, x_2, x_3, x_4) refers to a window packing that has x_1 genes from the first gene family, x_2 genes from the second gene family, x_3 genes from third gene family and x_4 genes from the fourth gene family.

In order to find all possible packings, we need to find all 4-tuples satisfying Equation (4); in this example $\sum_{j=1}^{4} x_j = 7$. Since j^{th} gene family can contribute x_j genes in $\binom{\phi_j}{x_j}$ ways, the 4-tuple (x_1, x_2, x_3, x_4) can contribute $\binom{\phi_1}{x_1}\binom{\phi_2}{x_2}\binom{\phi_3}{x_3}\binom{\phi_4}{x_4}$ window packings. For example, the tuple $(1, 1, 1, 4)$ can contribute $\binom{1}{1}\binom{2}{1}\binom{3}{1}\binom{4}{4} = 6$ window packings. Table 1 lists the set of all possible 4-tuples and the number of packings associated with each 4-tuple. By adding the number of packings for each 4-tuple, we get the total number of ways the window can be filled with genes from the four gene families as given in Equation (3). Here, $\mathcal{N}(7, 4, T) = 76$.

Table 1. Number of ways packing a window of size $w = 7$ with four gene families of size $\{1,2,3,4\}$

λ-tuple (x_1, x_2, x_3, x_4)	Number of packings
$(1, 1, 1, 4)$	6
$(1, 1, 2, 3)$	24
$(1, 1, 3, 2)$	12
$(1, 2, 1, 3)$	12
$(1, 2, 3, 1)$	4
$(1, 2, 2, 2)$	18
$\mathcal{N}(7, 4, T)$	76

3.2   Orthologous Clusters with Arbitrary Gene Families

We estimate the significance of a gene cluster using the probability that two windows, arbitrarily chosen from two random genomes, share at least m gene families. We enumerate over all sets of k gene families, for each value of k from m to r. For each such set, F, we determine the probability that W_1 contains only genes in families in F, including at least one from each family, followed by the conditional probability that at least l of the families in F also appear in W_2.

Expressed formally, the probability that W_1 and W_2 share *at least* m gene families is

$$q_o(m) = \sum_{k=m}^{r} \left[\sum_{F \in \mathcal{F}^k} p_1(F) \sum_{l=m}^{k} \sum_{\substack{E \in \mathcal{F}^l \\ E \subseteq F}} p_2(E) \right], \tag{5}$$

where $\mathcal{F}$ is the set of gene families in G_1 and G_2.

The probability that a given set, F, of k gene families is seen in W_1 is

$$p_1(F) = \binom{n_1}{r_1}^{-1} \mathcal{N}(r_1, k, F) \tag{6}$$

where $\mathcal{N}(r, k, F)$ is the number of window packings given by Equation (3). To determine, $p_2(E)$, we enumerate over all subsets of F of size l, where l ranges from m to k. For each subset, E, we seek the probability that each family in E is represented in W_2 at least once and that no other family in F appears in W_2. We exclude all other families in F to avoid overcounting.

At least l slots in W_2 must be filled with genes in E. The remaining $r_2 - l$ slots may be filled either from families in E or from families that do not appear in W_1; i.e., genes from $\mathcal{F} \setminus F$. Let z be the number of slots filled with genes from $\mathcal{F} \setminus F$. By considering all possible values of z, we obtain

$$p_2(E) = \binom{n_2}{r_2}^{-1} \sum_{z} \mathcal{N}(r_2 - z, l, E) \binom{n_2 - \Phi(F)}{z} \tag{7}$$

where $\Phi(F) = \sum_{j \in F} \phi_{2j}$. The parameter z ranges from $\max\{0, r_2 - \Phi(E)\}$ to $r_2 - l$ where $\Phi(E)$ is defined as above. The first term in the numerator is the number of ways of filling $r_2 - z$ slots with genes from the l families in E. The second term corresponds to all the ways of choosing the z outsiders from the set of genes not included in any gene family in W_1.

By substituting the expression in Equation (3) in Equations (6) and (7), we get a statistic for individual clusters in terms of n_1, r_1, n_2, r_2, m and the set of the gene families in G_1 and G_2. However, calculating this probability requires the enumeration of all subsets of k gene families. For each such subset, we must enumerate all packings satisfying Equation (4) and calculate a product of binomials for each packing. Computing this probability is prohibitively slow.

3.3   Orthologous Clusters with Fixed Size Families

The complexity of calculating $q(m)$ can be substantially reduced under the assumption that all gene families are of equal size, ϕ. When gene families are of equal size, it is not necessary to enumerate $\mathcal{F}^k$, since all subsets of k gene families are indistinguishable. We can simply replace the first term, $\sum_F p_1(F)$, in Equation (5) with the product of the number of sets of k gene families times $p_1(k)$, the probability that exactly k gene families of size ϕ are represented in the window:

$$\sum_{F \in \mathcal{F}^k} p_1(F) = \binom{|\mathcal{F}|}{k} p_1(k). \tag{8}$$

Invoking a similar transformation of the second term in Equation (5), the probability that W_1 and W_2 share at least m gene families simplifies to

$$q_o(m) = \sum_{k=m}^{r} \left[\binom{n_f}{k} p_1(k) \sum_{l=m}^{k} \binom{k}{l} p_2(l) \right].$$ (9)

Under the fixed size assumption, $p_1(k)$ and $p_2(l)$ correspond to the probability that exactly k families appear in W_1 and exactly l families appear in W_2, respectively. To calculate $p_1(k)$ and $p_2(l)$, we require an expression for $\mathcal{N}'(w, \lambda, \phi)$, the number of window packings when all families are of fixed size. When $\phi_j = \phi$, β_j reduces to $\binom{\phi}{x_j}$ and Equation (2) becomes

$$\alpha'(t) = \left[\binom{\phi}{1} t + \binom{\phi}{2} t^2 + \ldots + \binom{\phi}{\phi} t^\phi \right]^\lambda.$$ (10)

The number of ways of observing λ gene families in a window of size w is given by $[t^w] \alpha'(t)$, yielding

$$\mathcal{N}'(\omega, \lambda, \phi) = \sum_{(x_1, \cdots x_\lambda)} \binom{\phi}{x_1} \binom{\phi}{x_2} \cdots \binom{\phi}{x_\lambda},$$ (11)

where the sum is over the set of all λ-tuples $(x_1, \ldots, x_\lambda)$ satisfying Equation(4), under the constraint that $0 < x_i \leq \phi$, $\forall i$.

In this case, we can avoid enumerating the λ-tuples using the following simplification. Note that the right hand side of Equation (10) is a binomial series of the form $[(1 + t)^\phi - 1]^\lambda$. By applying two binomial expansions, we obtain

$$\alpha'(t) = (-1)^\lambda \sum_{i=0}^{\lambda} \left[(-1)^i \binom{\lambda}{i} \left(\sum_{j=0}^{i*\phi} \binom{i * \phi}{j} t^j \right) \right].$$ (12)

The number of ways of filling w slots with genes from the λ *fixed size* families is just $[t^w] \alpha'(t)$, yielding

$$\mathcal{N}'(w, \lambda, \phi) = (-1)^\lambda \sum_{i=0}^{\lambda} \left[(-1)^i \binom{\lambda}{i} \binom{i * \phi}{w} \right].$$ (13)

Notice that at least $\lceil \frac{w}{\phi} \rceil$ gene families are required to fill a window of size w. Substituting the expression for $\mathcal{N}'(w, \lambda, \phi)$ in Equation (6) and restricting the lower bound on the dummy variable i to $\lceil \frac{r_1}{\phi} \rceil$, we obtain

$$p_1(k) = \binom{n_1}{r_1}^{-1} (-1)^k \sum_{i=\lceil \frac{r_1}{\phi} \rceil}^{k} \left[(-1)^i \binom{k}{i} \binom{i * \phi}{r_1} \right].$$ (14)

Similarly, $p_2(l)$, the probability that W_2 contains exactly l gene families is

$$\binom{n_2}{r_2}^{-1} \sum_z (-1)^l \sum_{i=\lceil \frac{r_2-z}{\phi} \rceil}^{l} \left[(-1)^i \binom{l}{i} \binom{i * \phi}{r_2 - z} \right] \binom{n_2 - k\phi}{z}$$ (15)

where z ranges from $\max\{0, r_2 - k\phi\}$ to $r_2 - l$.

The fixed size approximation and the use of generating functions to enumerate window packings result in an efficient approximation to the probability that two windows, arbitrarily chosen from two random genomes, share at least m gene families. The general gene family model, Equations (6) and (7), requires implementation of an algorithm to enumerate all λ-tuples satisfying Equation (4). Furthermore, it is necessary to compute the product of λ binomial terms for each of the tuples in the enumeration. In contrast, Equations (14) and (15) require only a simple summation and can be easily computed in Mathematica. We can compute Equation (9) using the number of genes in each genome, the window sizes, gene family sizes and the number of gene families shared between the windows. Therefore, we only need information about the local regions and the aggregate properties of the genomes to determine significance of individual clusters.

4   Experimental Results

In this section, we demonstrate how the results derived in Section 3 can be applied to test the validity of a pair of putative homologous chromosomal regions. As an example, we applied our models, implemented in Mathematica, to the MHC-like region, so called because it contains a conserved linkage group that resides near the human Major Histocompatibility Complex. This conserved homologous region, which has four copies in mammalian genomes, has been discussed in the molecular evolution literature extensively [16–21, 30, 53].

In recent literature, there have been many papers about conserved linkage groups observed in eukaryotes that appear to be duplicated and, in some cases, also conserved across several distantly related species (surveyed in [30,31,53]). These include the mammalian MHC region, the regions surrounding the Hox clusters [22], a region on chromosome 8 in human (FGR) [23] and a region containing a Tbox subfamily on chromosomes 5 and 11 in mouse [28]. These clusters typically contain five to fifteen genes spread over a window of 15 to 300 slots. Most of these studies do not present any statistical analysis testing the significance of the clusters. A few use simple statistical tests based on a reference region model with no correction for gene family size [53,30,21].

Several of these conserved regions have been the focus of particular interest because four paralogous copies have been observed in mammals, leading to the speculation that they could have arisen through the early vertebrate tetraploidization postulated by Ohno [54]. The MHC-like region contains a conserved linkage group of roughly a dozen genes (depending on which analysis you look at) on chromosome 6p21.3 in human. Paralogous subsets of these genes are also found on human chromosomes 1, 9 and 19. The four putative paralogous regions in human and mouse have been studied extensively [16–20] as new sequence and mapping data has provided additional insights into the evolutionary implications of the regions. The increasing availability of whole sequence data has also led to the investigation of regions in other species with orthologous gene content and organization that is suggestive of common ancestry for the entire

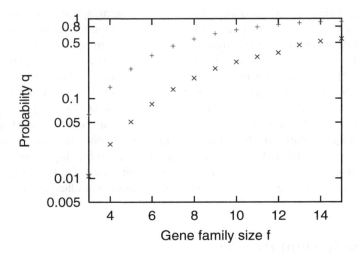

Fig. 2. Significance of MHC-like cluster using the fly genome as reference ($\times$) and the human genome as reference ($+$)

region. These include mouse, *C. elegans*, *D. melanogaster*, *S. Pombe* and several species of amphioxus [21,30,53].

We use a recent comparative analysis [30] of a chromosomal region on *Drosophila* chromosome X and the human MHC-like regions as an example for demonstrating the application of our statistical tests. Danchin *et. al.* [30] investigated a region delimited by *Drosophila* genes USP and Notch. USP is homologous to human RXRA, RXRB and RXRG. RXRA and Notch are "anchor" genes that bracket the MHCII region on human chromosome 6, a region also containing COL, ABC, RING and PSMB genes. Their analysis of *Drosophila* contigs in the public databases turned up 183 non-redundant transcripts in this region. Of these, 161 had significant matches in human, 32 of which included at least one significant hit in one or more of the MHC-like regions. Based on phylogenetic analysis, they [30] concluded that 19 of these were reliable orthologs. The two original anchor genes used to identify the region were eliminated from the study, since these were used to find the region do not constitute independent observations. Of the remaining 17 *Drosophila* genes in the study that had trusted human orthologs within one or more MHC-like regions, four fell into a region containing 44 genes on chromosome 6p21.3 in human.

We investigated the probability of observing such a cluster by chance using Equations (9), (14) and (15) and the following parameters $n_{hs} = 24194, n_{dm} = 13833, r_{hs} = 44, r_{dm} = 161$ and $m = 4$. Genome sizes were obtained from the *ensembl* database (*www.ensembl.org*). Note that our method to determine the significance of a gene cluster is not symmetric. When the sizes of the genomes and/or the windows are different ($n_1 \neq n_2$ and/or $r_1 \neq r_2$) the results will depend on which genome is designated G_1. Therefore, we estimated the cluster significance twice, using both the *Drosophila* and human regions as references.

The dependence of cluster statistics on gene family size is shown in Figure 2. These results show that cluster significance decreases rapidly with gene family size. Note that the probability of observing a gene cluster is slightly lower when *Drosophila* is used as reference. Since the second term in Equation (15) depends on n_2, the significance will decrease when G_2 is the larger genome. The probability of observing a homologous cluster structured like the human MHC-like cluster under the null hypothesis is greater than 0.1 for gene family sizes of four or greater and is close to one by the time ϕ reaches ten. While these numbers, taken alone, would suggest that the observed gene cluster is not statistically significant, a comprehensive analysis would require comparison of the chromosomal region in *Drosophila* with all four paralogous regions in human using a multiple testing approach. Our intent here is not to reanalyze the data or question the conclusions of the studies cited above, but rather to provide a concrete example of how our models can be put to practical use in real biological studies.

5   Conclusion

We have presented a new combinatorial approach to determine the significance of individual gene clusters. Our method takes gene family size into account and can be used to determine the significance of gene clusters in the absence of complete genomic context. We estimate the significance of gene clusters by determining the probability that two regions, containing r_1 and r_2 genes respectively, share at least m gene families. By using generating functions, we have developed tractable expressions for the estimating the probability of observing orthologous gene clusters in two genomes. To demonstrate the utility of the method, we have applied it to estimate the significance of a well-studied conserved region in the fly and human genome.

Acknowledgment

We thank P. Chebolu, A. Frieze, A. Goldman, R. A. Hoberman, N. Song and B. Vernot for helpful discussions and T. Kalra for nutritional support. This research was supported by NIH grant 1 K22 HG 02451-01 and a David and Lucille Packard Foundation fellowship.

References

1. Eichler, E.E., Sankoff, D.: Structural dynamics of eukaryotic chromosome evolution. Science **301** (2003) 793–797
2. Sankoff, D.: Rearrangements and chromosomal evolution. Curr Opin Genet Dev **13** (2003) 583–587
3. Sankoff, D., Nadeau, J.H.: Chromosome rearrangements in evolution: From gene order to genome sequence and back. PNAS **100** (2003) 11188–11189
4. Hurst, L.D., Pal, C., Lercher, M.J.: The evolutionary dynamics of eukaryotic gene order. Nat Rev Genet **5** (2004) 299–310

5. Tamames, J., Gonzalez-Moreno, M., Valencia, A., Vicente, M.: Bringing gene order into bacterial shape. Trends Genet **3** (2001) 124–126

6. Tamames, J.: Evolution of gene order conservation in prokaryotes. Genome Biol **6** (2001) 0020.1–0020.11

7. Blanchette, M., Kunisawa, T., Sankoff, D.: Gene order breakpoint evidence in animal mitochondrial phylogeny. J Mol Evol **49** (1999) 193–203

8. Cosner, M.E., Jansen, R.K., Moret, B.M.E., Raubeson, L.A., Wang, L.S., Warnow, T., Wyman, S.: An empirical comparison of phylogenetic methods on chloroplast gene order data in *Campanulaceae*. In Sankoff, D., Nadeau, J.H., eds.: Comparative Genomics. Kluwer Academic Press, Dordrecht, NL (2000) 99–121

9. Hannenhalli, S., Chappey, C., Koonin, E.V., Pevzner, P.A.: Genome sequence comparison and scenarios for gene rearrangements: A test case. Genomics **30** (1995) 299–311

10. Sankoff, D., Bryant, D., Deneault, M., Lang, B.F., Burger, G.: Early eukaryote evolution based on mitochondrial gene order breakpoints. J Comput Biol **3–4** (2000) 521–535

11. Sankoff, D., Deneault, M., Bryant, D., Lemieux, C., Turmel, M.: Chloroplast gene order and the divergence of plants and algae from the normalized number of induced breakpoints. In Sankoff, D., Nadeau, J.H., eds.: Comparative Genomics. Kluwer Academic Press, Dordrecht, NL (2000) 89–98

12. Chen, X., Su, Z., Dam, P., Palenik, B., Xu, Y., Jiang, T.: Operon prediction by comparative genomics: an application to the Synechococcus sp. WH8102 genome. Nucleic Acids Res **32** (2004) 2147–2157

13. Tamames, J., Casari, G., Ouzounis, C., Valencia, A.: Conserved clusters of functionally related genes in two bacterial genomes. J Mol Evol **44:** (1997) 66–73

14. Seoighe, C.: Turning the clock back on ancient genome duplication. Curr Opin Genet Dev **13** (2003) 636–643

15. Wolfe, K.: Yesterday's polyploids and the mystery of diploidization. Nature Rev Genet **2** (2001) 33–41

16. Endo, T., Imanishi, T., Gojobori, T., Inoko, H.: Evolutionary significance of intra-genome duplications on human chromosomes. Gene **205** (1997) 19–27

17. Hughes, A.L.: Phylogenetic tests of the hypothesis of block duplication of homologous genes on human chromosomes 6, 9, and 1. Mol Biol Evol **15** (1998) 854–70

18. Kasahara, M.: New insights into the genomic organization and origin of the major histocompatibility complex: role of chromosomal (genome) duplication in the emergence of the adaptive immune system. Hereditas **127** (1997) 59–65

19. Katsanis, N., Fitzgibbon, J., Fisher, E.: Paralogy mapping: identification of a region in the human MHC triplicated onto human chromosomes 1 and 9 allows the prediction and isolation of novel PBX and NOTCH loci. Genomics **35** (1996) 101–108

20. Smith, N.G.C., Knight, R., Hurst, L.D.: Vertebrate genome evolution: a slow shuffle or a big bang. BioEssays **21** (1999) 697–703

21. Trachtulec, Z., Forejt, J.: Synteny of orthologous genes conserved in mammals, snake, fly, nematode, and fission yeast. Mamm Genome **3** (2001) 227–231

22. Amores, A., Force, A., l. Yan, Y., Joly, L., Amemiya, C., Fritz, A., Ho, R., Langeland, J., Prince, V., Wang, Y.L., Westerfield, M., Ekker, M., Postlethwait, J.H.: Zebrafish hox clusters and vertebrate genome evolution. Science **282** (1998) 1711–1714

23. Spring, J.: Genome duplication strikes back. Nature Genetics **31** (2002) 128–129

24. Coulier, F., Pontarotti, P., Roubin, R., Hartung, H., Goldfarb, M., Birnbaum, D.: Of worms and men: An evolutionary perspective on the fibroblast growth factor (FGF) and FGF receptor families. J Mol Evol **44** (1997) 43–56

25. Lipovich, L., Lynch, E.D., Lee, M.K., King, M.C.: A novel sodium bicarbonate cotransporter-like gene in an ancient duplicated region: *SLC4A9* at 5q31. Genome Biol **2** (2001) 0011.1–0011.13

26. Lundin, L.G.: Evolution of the vertebrate genome as reflected in paralogous chromosomal regions in man and the house mouse. Genomics **16** (1993) 1–19

27. Pebusque, M.J., Coulier, F., Birnbaum, D., Pontarotti, P.: Ancient large-scale genome duplications: phylogenetic and linkage analyses shed light on chordate genome evolution. Mol Biol Evol **15** (1998) 1145–1159

28. Ruvinsky, I., Silver, L.M.: Newly indentified paralogous groups on mouse chromosomes 5 and 11 reveal the age of a T-box cluster duplication. Genomics **40** (1997) 262–266

29. Calabrese, P.P., Chakravarty, S., Vision, T.J.: Fast identification and statistical evaluation of segmental homologies in comparative maps. ISMB (Supplement of Bioinformatics) (2003) 74–80

30. Danchin, E.G.J., Abi-Rached, L., Gilles, A., Pontarotti, P.: Conservation of the MHC-like region throughout evolution. Immunogenetics **55** (2003) 141–8

31. Durand, D., Sankoff, D.: Tests for gene clustering. Journal of Computational Biology (2003) 453–482

32. Ehrlich, J., Sankoff, D., Nadeau, J.: Synteny conservation and chromosome rearrangements during mammalian evolution. Genetics **147** (1997) 289–296

33. Venter, J.C., et al.: The sequence of the human genome. Science **291** (2001) 1304–1351

34. Arabidopsis Genome Initiative: Analysis of the genome sequence of the flowering plant *Arabidopsis thaliana*. Nature **408** (2000) 796–815

35. Bansal, A.K.: An automated comparative analysis of 17 complete microbial genomes. Bioinformatics **15** (1999) 900–908

36. Bergeron, A., Corteel, S., Raffinot, M.: The algorithmic of gene teams. In Gusfield, D., Guigo, R., eds.: WABI. Volume 2452 of Lecture Notes in Computer Science. (2002) 464–476

37. Ponting, C.P., Schultz, J., Copley, R.R., Andrade, M.A., Bork, P.: Evolution of domain families. Adv Protein Chem **54** (2000) 185–244

38. Goldberg, L.A., Goldberg, P.W., Paterson, M.S., Pevzner, P., Sahinalp, S.C., Sweedyk, E.: The complexity of gene placement. Journal of Algorithms **41** (2001) 225–243

39. Heber, S., Stoye, J.: Algorithms for finding gene clusters. In: WABI. Volume 2149 of Lecture Notes in Computer Science. (2001) 254–265

40. Heber, S., Stoye, J.: Finding all common intervals of k permutations. In: Proceedings of CPM01. Volume 2089 of Lecture Notes in Computer Science. (2001) 207–218

41. Nadeau, J., Sankoff, D.: Counting on comparative maps. Trends Genet **14** (1998) 495–501

42. O'Brien, S.J., Wienberg, J., Lyons, L.A.: Comparative genomics: lessons from cats. Trends Genet **10** (1997) 393–399

43. Overbeek, R., Fonstein, M., D'Souza, M., Pusch, G.D., Maltsev, N.: The use of gene clusters to infer functional coupling. Proc Natl Acad Sci U S A **96** (1999) 2896–2901

44. Wolfe, K.H., Shields, D.C.: Molecular evidence for an ancient duplication of the entire yeast genome. Nature **387** (1997) 708–713

45. Hoberman, R., Durand, D.: Incompatible desiderata of gene cluster properties. "Proceedings of the 3rd RECOMB Satellite Workshop on Comparative Genomics", Lecture Notes in Bioinformatics, Springer Verlag, in press. (2005)

46. Hoberman, R., Sankoff, D., Durand, D.: The statistical analysis of spatially clustered genes under the maximum gap criterion. Journal of Computational Biology, in press (2005)

47. Li, Q., Lee, B.T.K., Zhang, L.: Genome-scale analysis of positional clustering of mouse testis-specific genes. BMC Genomics **6** (2005) 7

48. Adams, M.D., et al.: The genome sequence of *Drosophila melanogaster*. Science **287** (2000) 2185–2195

49. Huynen, M.A., Bork, P.: Measuring genome evolution. PNAS **95** (1998) 5849–5856

50. Tatusov, R.L., Koonin, E.V., Lipman, D.: A genomic perspective on protein families. Science **278** (1997) 631–637

51. Manning, G., Whyte, D.B., Martinez, R., Hunter, T., Sudarsanam, S.: The protein kinase complement of the human genome. Science **298** (2002) 1912–1934

52. Polya, G.: Notes on introductory combinatorics. Birkhauser (1983)

53. Abi-Rached, L., Gilles, A., Shiina, T., Pontarotti, P., Inoko, H.: Evidence of en bloc duplication in vertebrate genomes. Nat Genet **31** (2002) 100–105

54. Ohno, S.: Evolution by genome duplication. Berlin: Springer Verlag (1970)

Power Boosts for Cluster Tests

David Sankoff and Lani Haque

Department of Mathematics and Statistics, University of Ottawa,
585 King Edward Street, Ottawa, Canada, K1S 4T8
{sankoff, lhaque}@uottawa.ca

Abstract. Gene cluster significance tests that are based on the number of genes in a cluster in two genomes, and how compactly they are distributed, but not their order, may be made more powerful by the addition of a test component that focuses solely on the similarity of the ordering of the common genes in the clusters in the two genomes. Here we suggest four such tests, compare them, and investigate one of them, the maximum adjacency disruption criterion, in some detail, analytically and through simulation.

1   Introduction

The detection of a number of genes in close proximity in two genomes may suggest an evolutionary association of these genes or indicate a functional relationship among them. Recently studied tests of significance of such gene clusters [2–4] have focused on the number of genes in common in relatively short chromosomal intervals in the two genomes, and not on the order of these genes. For different gene clusters having the same number of genes, spatially distributed in the same way, we might want to consider those clusters where the order is more similar, though not necessarily identical, in the two genomes as being more significant, and more indicative of a historical or functional relationship, than clusters whose gene orders in the two genomes bear little relationship. For tests such as those in the above-cited studies, where the significance level is independent of gene order within the cluster, this level can be enhanced by taking into account the significance of the gene-order similarity within the cluster in the two genomes. This combined test would have greater power against the null hypothesis of random gene order at the genomic scale in favour of alternative hypotheses derived from evolutionary or functional models.

In this this paper, we first suggest four different ways of defining an order-based "boost" to cluster significance, and investigate one of them in enough detail so that it can be used for all cluster sizes and all values of the similarity criterion. In Section 2 we define four measures of gene-order divergence and link them to previous work on genome rearrangement and gene clusters. In Section 3, we show some properties of the maximum adjacency disruption measure of order similarity. In the rest of the paper we develop tests based on this measure; in Section 4, an exact test for certain values of the maximum difference measure, and in Section 5 exact tests for all small clusters. For large clusters,

A. McLysaght et al. (Eds.): RECOMB 2005 Ws on Comparative Genomics, LNBI 3678, pp. 121–130, 2005.

we tabulate significance levels based on large scale simulations in Section 6, and on a simplified probabilistic model in Section 7. In Section 8, we compare the critical regions three of the four proposed tests, on moderate-sized clusters. In Section 9, we discuss the applicability of our method, and directions for further research.

2   Four Measures of Gene-Order Similarity and Their Motivations

Suppose we have identified, using some existing criterion, e.g., one of those in [2–4], k genes that form a cluster in both genome A and genome B. Number the clustered genes in genome A in order from 1 to k (ignoring any intervening genes that are not in the scope of the cluster in genome B) and let $g_1, \cdots, g_k$ be the order of these same genes in genome B. Similarly, re-number the genes from 1 to k according to their order on genome B, and let $h_1, \cdots, h_k$ be the order of these same genes in genome A.

1. The maximum adjacency disruption criterion (MAD):

$$\text{MAD} = \max_{i=1,\cdots,k-1} \{\max\{|g_i - g_{i+1}|, |h_i - h_{i+1}|\}\},$$

   the maximum, over all pairs of adjacent genes in the cluster in either genome, of the difference in their positions in the gene order in the cluster in the other genome. A low value of MAD means that no gene in the cluster has drifted far from its position in the ancestral genome. MAD is symmetric with respect to A and B. An asymmetric criterion somewhat similar to MAD was used in [1].

2. The summed adjacency disruption criterion (SAD):

$$\text{SAD} = \sum_{i=1,\cdots,k-1} \{|g_i - g_{i+1}| + |h_i - h_{i+1}|\},$$

   the sum, over all pairs of adjacent genes in the cluster in both genomes, of the difference in their positions in the gene order in the cluster in the other genome. This measures the overall movement of genes within the cluster from their positions in the ancestral genome.

3. The breakpoint metric (BAD):

$$\text{BAD} = \#_{(i=1,\cdots,k-1)}\{|g_i - g_{i+1}| > 1\},$$

   the number of times a pair of genes adjacent in the cluster in one genome is not adjacent in the other. This is in effect a simple count of the adjacency disruptions and has been used in comparisons of entire gene orders of genomes [5], whereas here we are focusing on the order of genes in the cluster only.

4. The rearrangement distance (RAD). The number of rearrangement operations (e.g., inversions, transpositions, block interchanges) required to transform the order of the genes in one cluster into the order in the other one [6]. This is the only measure we do not analyze here, for reasons detailed in Section 9.

To test for a significant level of gene-order correspondence between a cluster's realizations in two genomes, we need to know the distribution of the test statistic under a suitable null hypothesis, normally that gene order is purely random. This may be done by counting the number of permutations of the integers from 1 to k having a given value of the statistic, either exhaustively, or by means of a computing formula if this is available, or estimated through simulation or approximated by a continuous model. We will calculate the distribution of the MAD statistic, using all these approaches, depending on the cluster size k, and we will compare it to SAD and BAD on all permutations of size 11 and 12.

3   Maximum Adjacency Disruption; One-Sided and Two-Sided

Defining the one-sided versions of MAD,

$$\mathrm{MAD}_{AB} = \max_{i=1,\cdots,k-1}\{|g_i - g_{i+1}|\}, \mathrm{MAD}_{BA} = \max_{i=1,\cdots,k-1}\{|h_i - h_{i+1}|\},$$

it is not the case that MAD_{AB} always equals MAD_{BA}.

Example 1. : Consider B = 3 1 2 4. Then $\mathrm{MAD}_{AB} = 2$. Renumbering the genes in order in B as 1 2 3 4 translates into A = 2 3 1 4, so that $\mathrm{MAD}_{BA} = 3$.

Limiting the MAD value of a cluster not only constrains pairs of adjacent genes in one genome from being far apart in the other, it also keeps genes in the larger neighbourhood of a given gene from dispersing too widely. For a given k,

Proposition 1. *If*

$$|g_i - g_{i+1}| \le a, \ for \ 1 \le i < k$$

then

$$|g_i - g_j| \le a|i - j|, \ for \ 1 \le i, j \le k.$$

Proof:

$$|g_i - g_j| = |\sum_{m=i}^{j-1} g_m - g_{m+1}| \tag{1}$$

$$\le \sum_{m=i}^{j-1} |g_m - g_{m+1}| \tag{2}$$

$$\le a|i - j| \tag{3}$$

Corollary 1. *Let $f(x)$ be an increasing concave function on $[1, \cdots, k]$. Then*

$$f(|i - j|) \le f(1)|i - j|.$$

4   Enumerating Clusters Where $a = 1, 2$ and 3, and Where $a = k - 2$ and $a = k - 1$

Being able to calculate the number of clusters of size k with a specific value of a criterion is helpful for the construction of tests based on this criterion. Computing formulae are currently known only for BAD (entry A001100 in Sloane's On-Line Encyclopedia of Integer Sequences [7].) Here we give partial results for MAD. Let $n(a, k)$ be the number of clusters of length k where MAD $\leq a$.

Proposition 2. *The following statements hold:*

1. $n(1, k) = 2$, *for all* k
2. *For* $a = 2$, $k > 5$

$$n(2, k) = n(2, k - 1) + n(2, k - 3).$$

3. *For* $a = 3$, $k > 12$

$$n(3, k) = n(3, k - 1) + n(3, k - 2) + 3n(3, k - 4)$$
$$+3n(3, k - 5) - 3n(3, k - 6) - 3n(3, k - 7)$$

4. $n(k - 1, k) = k!$, *for* $k > 1$.
5. $n(k - 2, k) = (k - 2)!(k^2 - 5k + 8)$, *for* $k > 2$.

Proof:

1) The only clusters that have MAD $= 1$ are $1 \cdots k$ and $k \cdots 1$.

2) One type of "valid" cluster, i.e., one that has MAD ≤ 2, is of form $k \, \gamma$, where γ is any valid cluster on the first $k - 1$ integers starting with $k - 1$ or $k - 2$, or of form $\kappa \, k$, where κ is any valid cluster on the first $k - 1$ integers ending with $k - 1$ or $k - 2$. There are $n(2, k - 1)$ of these. The other type of cluster that has MAD ≤ 2 is of form $k - 1 \, k \, k - 2 \, \gamma$, where γ is any valid cluster on the first $k - 3$ integers starting with $k - 3$ or $k - 4$, or of form $\kappa \, k - 2 \, k \, k - 1$, where κ is any valid cluster on the first $k - 3$ integers ending with $k - 3$ or $k - 4$. There are $n(2, k - 3)$ of these. Thus, for large enough k (starting with $k = 6$), we have $n(2, k) = n(2, k - 1) + n(2, k - 3)$.

3) An argument similar to that for $a = 2$, but with many more cases, shows that for k even,

$$n(3, k - 2) = n(3, k - 3) + n(3, k - 5) + 3n(3, k - 6) + 3n(3, k - 7)$$
$$+ \sum_{i=2}^{\frac{k-4}{2}} n(3, k - (2i + 3))$$

So that

$$n(3, k) - n(3, k - 2)$$
$$= n(3, k - 1) + 3n(3, k - 4) + 3n(3, k - 5) - 3n(3, k - 6) - 3n(3, k - 7)$$

and the analogous calculation for k odd yields the same result. Therefore,

$$n(3, k) = n(3, k - 1) + n(3, k - 2)$$
$$+3[n(3, k - 4) + n(3, k - 5) - n(3, k - 6) - n(3, k - 7)].$$

4) The criterion $a = k - 1$ holds for any of the $k!$ clusters.

5) This follows from $n(k-2, k) = n(k-1, k) - n'(k-1, k)$, where $n'(k-1, k) = 4(k - 2)(k - 2)!$ is the number of clusters with a maximum disruption score of exactly $k-1$. The latter is the number of clusters such that $\{g_1, g_k\} = \{1, k\}$ plus the number such that $\{h_1, h_k\} = \{1, k\}$ less the number where both conditions hold. All of these may be evaluated in a straightforward manner, yielding the above expression for $n'(k - 1, k)$.

The first few values of $n(2, k)$ are $0, 0, 6, 8, 12, 18, 26, \cdots$. The recurrence $n(2, k) = n(2, k-1) + n(2, k-3)$ only "kicks in" at $k = 6$. The particular series of numbers starting with $0, 0, 6, 8, 12, 18$ has not been mentioned before, but the recurrence, with different initial values, has cropped up in many different contexts, listed as sequence A000930 in Sloane's On-Line Encyclopedia of Integer Sequences [7].

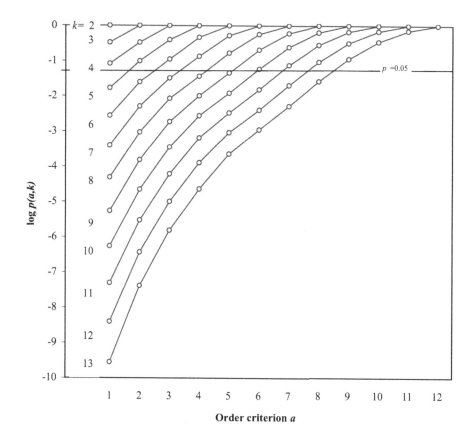

Fig. 1. Proportion of clusters of size k with maximum adjacency disruption $\leq a$; exact counts for small k

5   Exact Enumeration for Moderate Values of k

As a first step in constructing a usable test based on MAD, for $k \leq 13$, we
generated all permutations on the integers from 1 to k and calculated the MAD
value for each when compared to the identity permutation. We then calculated
the $n(a, k)$ as defined in Section 4, normalized by $k!$ to compute the p-value
$p(a, k)$, and plotted the results on a logarithmic scale in Figure 1. Given a cluster
with MAD $= a$, then, its statistical significance can be assessed from the curve
for clusters of size k.

6   Simulations for Large k

Though it would be feasible using high-performance methods to calculate the
test exactly for k somewhat larger than 13, this would exceed 16 or 17 with
great difficulty. Instead, we simply constructed 100,000 random clusters with

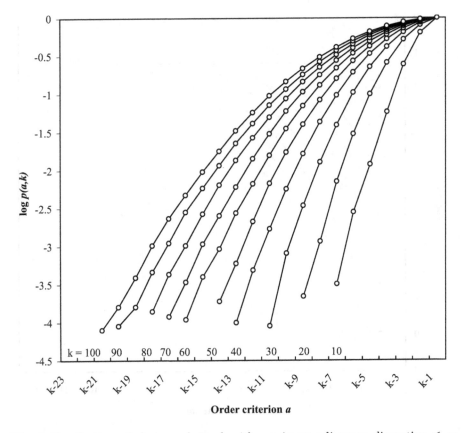

Fig. 2. Proplortion of clusters of size k with maximum adjacency disruption $\leq a$;
estimated values for larger k based on 100,000 randomly generated clusters

k terms, for k running from 10 to 100, in steps of 10. The curves in Figure 2 were constructed in the same way as the previous figure, except that the normalization was by 100,000 instead of by $k!$, and that the curves are plotted against $a = k - 1, k - 2, \cdots$ rather than against $a = 1, 2, \cdots$.

7   A Model for Large k

Let $\alpha = \frac{a}{k}$. To model the MAD criterion $|g_i - g_{i+1}| \le a$ for large k under the null hypothesis, i.e., for genes randomly ordered within clusters, we consider two points at random on the real unit interval, and ask what is the probability they are within α of each other, where $0 < \alpha < 1$. This is just $1 - (1-\alpha)^2$, representing the probability of adjacency disruption of less than α. Since k is large, we explore the assumption that these disruptions are independent across all $2k$ adjacencies in the two genomes. Then the probability that all the disruptions are less than α is $[1 - (1-\alpha)^2]^{2k} = [\alpha(2-\alpha)]^{2k}$.

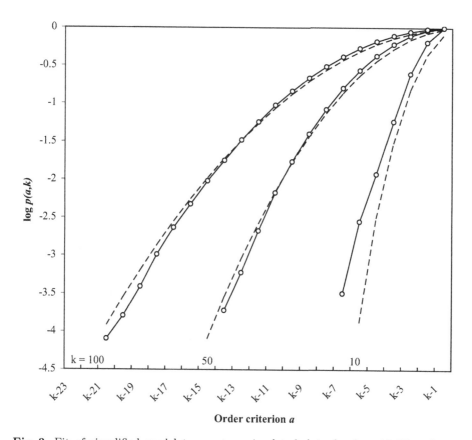

Fig. 3. Fit of simplified model to exact or simulated data for $k = 10, 50$ and 100. Dashed lines represent model, lines with data points drawn from exact values ($k = 10$) or simulations ($k = 50, 100$).

Plotting $p = [\frac{a}{k}(2 - \frac{a}{k})]^{2k}$ in Figure 3 demonstrates an improved fit of this simplified model for large values of k (50-100), compared with the case $k = 10$.

8   A Comparison of Adjacency Disruption Measures

Tests based on the four measures listed in Section 2 will not, of course, all have the same critical region. To compare the MAD, BAD and SAD criteria, we calculated them on all 4×10^7 clusters of size 11, and counted how many clusters fell into the critical regions of both members of a pair of tests. We aimed for equal critical regions of size close to 5%, but because we were restricted to discrete choices of a, the closest we could come was defined by an a for each test (detailed in the table legend) that resulted in approximately 8% of the criteria values falling on or below this threshold. We then repeated this for the 4.79×10^8 clusters of size 12; this time the critical regions closest to 5% all had size approximately 2%.

Table 1. Differences among critical regions of three tests. $k = 11$, total clusters 4×10^7, criteria for MAD: $a \leq 7$, BAD: $a \leq 6$, SAD: $a \leq 63$. $k = 12$, total clusters 4.79×10^8, MAD: $a \leq 7$, BAD: $a \leq 6$, SAD: $a \leq 67$.

Clusters in critical regions ($\alpha = 7.5 - 9.0\%$)	MAD	SAD	BAD
$(k = 11) R_i (\times 10^6)$	3.01	3.06	3.55
	Compared with SAD	BAD	MAD
Intersection $R_i \cap R_j$	1.09	1.7	0.51
Union $R_i \cup R_j$	4.98	4.91	6.05
Symmetric difference $R_i \Delta R_j$	3.89	3.2	5.54
Normalized difference $\frac{R_i \Delta R_j}{R_i \cup R_j}$	**0.782**	**0.653**	**0.916**
Clusters in critical regions ($\alpha = 2.0 - 2.3\%$)	MAD	SAD	BAD
$(k = 12) R_i (\times 10^6)$	9.65	10.1	11
	Compared with SAD	BAD	MAD
Intersection $R_i \cap R_j$	3.28	3.91	0.81
Union $R_i \cup R_j$	16.46	17.18	19.84
Symmetric difference $R_i \Delta R_j$	13.18	13.27	19.03
Normalized difference $\frac{R_i \Delta R_j}{R_i \cup R_j}$	**0.801**	**0.772**	**0.959**

Table 1 shows that SAD is closer to both MAD and BAD than they are to each other. This is understandable in that the sum of the adjacency disruptions should reflect not only the number of disruptions but also the size of the largest one. The number of disruptions (BAD), however, and the size of the largest one (MAD), would seem only very indirectly related. Indeed, almost all the clusters satisfying both the MAD and BAD criteria (0.51×10^6 for the $k = 11$ experiment, and 0.81×10^6 for the $k = 12$ experiment) also satisfy the SAD criterion (0.48×10^6 and 0.75×10^6, respectively).

9 Discussion

The study of gene clusters is increasingly focused on genomes where the genes have been located by virtue of genomic sequencing. Such clusters include not only gene position and hence gene order, but also gene orientation, or strandedness. In comparative genomics, such data are represented not by ordinary permutations, but by signed permutations, where the sign on a term in genome B indicates the DNA strand, or reading direction, relative to the direction of the same gene in genome A. In this context, we can simply ignore the sign, or else devise some way of taking it into account. For BAD, and particularly for RAD, signed permutations are the natural domain of application. This is one reason why we did not analyze RAD in the present study. The main adaptation is that in signed permutations, the configuration $i + 1, i$ in genome B is considered a disruption of adjacency, but $-(i + 1), -i$ is not, when the two genes are ordered as $i, i + 1$ in genome A. For MAD and SAD, there is no natural way of taking sign into account and it seems most appropriate to ignore it.

Note that both MAD and SAD are asymmetrical, in that MAD_{AB} is not always equal to MAD_{BA} and SAD_{AB} is not always equal to SAD_{BA}. MAD is symmetrical by virtue of taking the maximum over both directions, and SAD by summing over both directions. Both BAD and RAD are symmetrical, on the other hand, and that is why it suffices to define them asymmetrically as we did in Section 2.

How should our tests for gene order be combined with tests for gene clustering such as r-windows and max-gap in [2–4]? The intuitive notion of a cluster involves spatial proximity of a group of genes similarly ordered in both genomes. But the most straightforward way of combining the two kinds of test, simply multiplying the two significance levels, is not an acceptable strategy, since it then only requires that one of the two tests be significant for the combined test to be significant. For example, if a putative cluster with k genes is evenly spaced across the entire genome, so that it is really the antithesis of the intuitive notion a cluster, but the order is $1 \cdots k$, then it will be still be highly significant (for large k) when the two significance levels (clustering and order) are multiplied together. The critical region of the combined test thus includes groups of genes which cannot be considered clusters. This problem does not seem to have been studied in the literature, where it is sometimes assumed that the significance level of a cluster of k genes may be enhanced by a factor of $(k!)^{-1}$ if the gene order is identical in both genomes.

For clusters with borderline (or better) p-values, however, the multiplicative strategy effectively boosts the power of the cluster test against evolutionarily or functionally meaningful alternatives.

Acknowledgements

Research supported in part by grants from the Natural Sciences and Engineering Research Council of Canada (NSERC). DS holds the Canada Research Chair in

Mathematical Genomics and is a Fellow of the Evolutionary Biology Program of the Canadian Institute for Advanced Research. The authors thank the referees for numerous corrections and helpful comments.

References

1. Calabrese, P.P., Chakravarty, S. and Vision, T.J. 2003. Fast identification and statistical evaluation of segmental homologies in comparative maps. Bioinformatics 19, i74–i80.
2. Durand, D. and Sankoff, D., 2003. Tests for gene clustering. Journal of Computational Biology 10, 453–482.
3. Hoberman, R., Sankoff, D. and Durand, D. 2005. The statistical significance of max-gap clusters. in Lagergren, J. (ed) RECOMB Satellite Workshop on Comparative Genomics. LNBI 3388, 55–71. Berlin, Heidelberg: Springer Verlag.
4. Hoberman, R., Sankoff, D. and Durand, D. 2005. The statistical analysis of spatially clustered genes under the maximum gap criterion. Journal of Computational Biology, in press.
5. Sankoff, D. and .Blanchette, M. 1998. Multiple genome rearrangement and breakpoint phylogeny. Journal of Computational Biology 5, 555-570
6. Sankoff, D. and El-Mabrouk, N. 2002. Genome rearrangement. in Jiang, T., Smith, T., Xu, Y. and Zhang, M (eds) Current Topics in Computational Biology, 135–155. Cambridge, MA: MIT Press.
7. N. J. A. Sloane. 2005. The On-Line Encyclopedia of Integer Sequences, http://www.research.att.com/~njas/sequences/.

Reversals of Fortune

David Sankoff[1], Chungfang Zheng[2], and Aleksander Lenert[3]

[1] Department of Mathematics and Statistics
[2] Department of Biology
[3] Program in Biopharmaceutical Science,
University of Ottawa, Canada
{sankoff, czhen033, alene096}@uottawa.ca

Abstract. The objective function of the genome rearrangement problems allows the integration of other genome-level problems so that they may be solved simultaneously. Three examples, all of which are hard: 1) Orientation assignment for unsigned genomes. 2) Ortholog identification in the presence of multiple copies of genes. 3) Linearisation of partially ordered genomes. The comparison of traditional genetic maps by rearrangement algorithms poses all these problems. We combine heuristics for the first two problems with an exact algorithm for the third to solve a moderate-sized instance comparing maps of cereal genomes.

1 Introduction

The first chromosomal map dates from 1913 [30], at the same time the definitive chromosomal theory of heredity [19] was being elaborated. Soon comparative mapping had become an integral part of genetic research, e.g., Fig. 1 in [31], published in 1921. Long before the genomic era, comparative maps existed for *Drosophila* and other insects, mammals, including humans, livestock and rodents, cereals and other cultivars and other eukaryotic and prokaryotic groups.

Despite their immediate availability and the wealth of evidence they contain about evolutionary history, traditional comparative maps were bypassed when genome rearrangement algorithms ([14,15]), inspired by analyses of organelle and other small genomes (e.g., [23,29]), were adapted for direct use on DNA segments derived from whole nuclear genome sequences [24,2,4].

In this paper we discuss an approach to the application of rearrangement methods to traditional comparative maps, i.e., maps based on estimates of gene and marker locations in nuclear genome, and not directly on genome sequence. First, what are the difficulties we encounter when we attempt this?

Coarseness. Lack of resolution of the maps, i.e., two or more genes being mapped to the same position in one of the genomes. Genome rearrangement algorithms require that the input markers be totally ordered along each chromosome.

Missing Data. Order ambiguity in composite maps. Because maps constructed from a single type of experimental data usually contain a limited number of markers, we are motivated to combine maps for the same genome from

A. McLysaght et al. (Eds.): RECOMB 2005 Ws on Comparative Genomics, LNBI 3678, pp. 131–141, 2005.

different sources. Two genes or markers which are not ordered by any of the component maps will often remain unordered in the composite map. Again, rearrangement algorithms require that the input markers be totally ordered along each chromosome.

No Signs. No information about reading direction, i.e., which DNA strand the gene or marker is on. This information is not available from many of the methods used to construct maps. Genome rearrangement algorithms require this "orientation" information for efficient and exact execution.

Matches. Uncertain orthology.

> **Notation.** Different nomenclatural traditions in the genetics communities producing the chromosomal maps for two species mean different annotations and difficulties for the analyst in deciding which markers in one genome correspond to markers in the other. Rearrangement algorithms require that genes or other markers on the two genomes be unequivocally paired as being derived from a single entity in an ancestral genome.

> **Paralogy.** Several copies or near copies of a gene in a map. This leads to a one-to-many or many-to-may correspondence between the two maps. Genome rearrangement algorithms require one-to-one correspondences as input.

Conflicts. When two or more relatively sparse maps of a genome, compiled from different sources, are combined prior to comparison with the map of another genomes, there is often conflict concerning the orders of a some of the markers on both maps.

With the possible exception of **paralogy**, these difficulties are neatly avoided when complete genome sequences are being compared at the sequence level [24,2,4], though of course there are many other technical problems to be solved in that approach.

The difficulties listed above all have in common that we are missing some essential biological information required to carry out genome rearrangement analysis. Moreover, in each case (except **notation**) THE GENOME REARRANGEMENT PROBLEM MAY BE REFORMULATED IN SUCH A WAY THAT THE SOLUTION NOT ONLY PROVIDES A MINIMAL SERIES OF REVERSALS AND/OR TRANSLOCATIONS NECESSARY TO TRANSFORM ONE GENOME INTO ANOTHER, BUT ALSO SUPPLIES AN OPTIMAL ESTIMATE OF THE MISSING INFORMATION. It is the comparative context, together with the rearrangement-minimizing objective function, which "fills in" the gaps in our biological knowledge in the most reasonable way. This unexpected bounty from the rearrangement analysis is what is alluded to in the title of this paper.

Exact algorithms have been published to take care of **coarseness, missing data, no sign** and **paralogy**, all requiring exponential worst-case computing time. The latter two, the topics of Sections 3 and 4, respectively, have been proved NP-hard, and we have conjectured as much for the first two, which are the main focus of this paper, as discussed in Section 5. As for the **notation** problem, we may rely on one of the curated comparative browsers, such as Gramene [36] for cereals and some other plants, the NCBI Human-Mouse homology maps [20], UCSC Genome browser [35] for animals, or CompLDB [21] for livestock.

Solution of a typical comparative map rearrangement problem would require treating at least **coarseness, no sign** and **paralogy** simultaneously, and usually **missing data** and **conflict** as well. We will state the pertinent combinatorial optimization problem, but its exact solution would be feasible only on very particular, small instances. We do, however, give results of applying an exact algorithm allowing for **coarseness** and **missing data**, in all generality, applied to data where **no sign, paralogy** and **conflict** are dealt with heuristically during preprocessing, using some of the key ideas in their respective algorithms.

In Section 2, however, we will start with the essentials of genome rearrangement theory.

2   The Bicoloured Graph in Rearrangement Algorithms

Hannenhalli and Pevzner [15] showed how to find a shortest sequence of reversals and translocations that transform one completely specified genome χ with n genes on k chromosomes into another genome ψ of the same size but with h chromosomes, in polynomial time. Completely specified means that each chromosome is totally ordered, the sign of each gene is known, and there is no paralogy.

As described in [34], we construct a bicoloured graph on $2n + 2k$ vertices that decomposes uniquely into a set of alternating-coloured cycles and $h + k$ alternating-colour paths. First, each gene x in χ determines two vertices, x_t and x_h. Two dummy vertices e_{i_1} and e_{i_2} are added to the ends of each chromosome χ_i. The adjacencies in χ determine red edges. If x is the left neighbour of y in χ, and both have positive polarity, then x_h is connected by a red edge to y_t. If they both are negative, x_t is joined to y_h. If x is positive and y negative, or x is negative and y positive, x_h is joined to y_h, or x_t is joined to y_t, respectively. If x is the first gene in χ_i, then e_{i_1} is joined to x_t or x_h depending on whether x has positive or negative polarity, respectively. If x is the last gene, then e_{i_2} is joined to x_t or x_h depending on whether x is negative or positive.

Black edges are added according to the same rules, based on the adjacencies in genome ψ, though no dummy vertices are added in this genome.

Each vertex is incident to exactly one red and one black edge edge, except for the dummies in χ and the (non-dummy) vertices at the ends of chromosomes in ψ, which are each incident to only a red edge. The bicoloured graph decomposes uniquely into a number of alternating cycles plus $h + k$ alternating paths terminating in either the dummy vertices of χ or the end vertices of ψ, or one of each. Suppose the number of these paths that terminate in at least one dummy vertex is $j \leq h + k$. If the number of cycles is c, then the minimum number of reversals r and translocations t necessary to convert χ into ψ is given by:

$$r + t = n - j - c + \theta \tag{1}$$

where θ is a correction term that is usually zero for simulated or empirical data. For simplicity of exposition, we ignore this correction here. Indeed, in a recent framework [11] allowing p transpositions and more general block interchanges via circular intermediate chromosomal fragments, $\theta \equiv 0$, and we simply have

$$r + t + 2p = n - j - c. \tag{2}$$

The $n - j - c$ actual rearrangement steps for transforming χ into ψ can then be found via certain well-defined operations on the cycles of the bicoloured graph.

3   Sign Assignment

Our first problem is that of adding signs to an unsigned genome so as to a achieve a minimal reversal distance to the identity permutation $1, \cdots, n$. This is equivalent to the problem of sorting an unsigned permutation, known to be NP-hard [7].

As conjectured in [17] and proved in [16], for all segments of the permutation consisting of three or more consecutive integers (strips) in increasing order, plus signs can be given to all these integers, and for all decreasing strips, minus signs can be given, and this assignment is consistent. with a solution. In [16], it is also shown how to give signs to 2-strips. The algorithm these authors develop is exponential only in s, the number of singletons, and is polynomial if s is $O(\log n)$. Unfortunately, in comparative maps s often seems closer to $O(n)$.

Though there is much recent literature on approximation to unsigned reversal distance, relatively little work has been done on exact algorithms. Caprara *et al.* [8] have implemented a branch-and-price algorithm that enables the rapid sorting of up to 200 elements. Tesler (personal communication) has extended the approach in [16] to reversal and translocation distance, and implemented it in GRIMM [33].

4   Duplicate Genes, Paralogy, Gene Families

When there are paralogs, gene orders cannot be modeled as permutations, but only as more general strings. Though sorting strings by reversals can be done in polynomial time, this does not automatically give the reversal distance between strings, in contrast to sorting permutations by reversals, which is equivalent to calculating reversal distance. Indeed, reversal distance for strings is NP-hard [26].

The problem in analysing genomes containing paralogs is how to decide which paralog in one genome should be identified with which one in the other genome, in a biologically meaningful way. Thus string-based analyses that attempt to match all or as many as possible of the paralogs of a gene in one genome to distinct paralogs in the other are only meaningful under the often questionable assumption that all paralogs were present in the common ancestor genome.

A less ambitious, but biologically more reasonable, approach is to try to match only one paralog of each gene in one genome to one in the other, such that the gene orders of the matched paralogs (the *exemplars*) of each family result in a minimal reversal distance [27]. The hypothesis motivating this is that genomes reduced to contain only these exemplars will better tend to reflect the actual reversal history than reduced genomes made up of any other choice of exemplars, using a parsimony argument.

There is a growing literature on the problem of incorporating paralogy into genome rearrangement theory. This is most meaningfully carried out within the phylogenetic context [28,3], taking into account that the origin of paralogs in duplication events may occur on earlier or later branches of the evolutionary tree. In addition to work characterizing, approximating or generalizing the exemplar approach [6,22], there is research on rearrangement in the context of string theory [10,26], conserved interval/block theory [1,3] and other a number of other approaches [9,32]. Virtually all of these are based on the same principle, that matching of paralogs should minimize the rearrangement distance

5   Partial Order

A linear map of a chromosome that has several genes or markers at the same position π, because their order has not been resolved, can be reformulated as a partial order, where all the genes before π are ordered before all the genes at π and all the genes at π are ordered before all the genes following π, but the genes at π are not ordered amongst themselves.

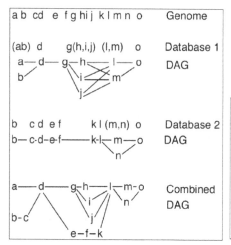 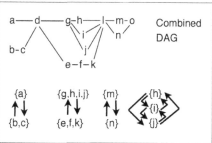

Fig. 1. (Left) Construction of DAGs from individual databases each containing partial information on genome, due to missing genes and missing order information, followed by construction of combined DAG representing all known information on the genome. All edges directed from left to right. (Right) Edges added to DAG to obtain DG containing all linearization as paths (though not all paths in the DG are linearizations of the DAG!). Each arrow represents a set of directed edges, one from each element in one set to each element of the other set.

For genomes with two or more gene maps constructed from different kinds of data or using different methodologies, there is only one meaningful way of combining the order information on two (partially ordered) maps of the same chromosome containing different subsets of genes. Assuming there are no conflicting order relations ($a < b, b < a$) nor conflicting assignments of genes to chromosomes among the data sets, for each chromosome we simply take the union of the partial orders, and extend this set through transitivity.

All the partial order data on a chromosome can be represented in a directed acyclic graph (DAG) whose vertex set is the union of all gene sets on that chromosome in the contributing data sets, and whose edges correspond to just those order relations that cannot be derived from other order relations by transitivity. The outcome of this construction is illustrated on the left of Figure 1.

We can extend genome rearrangement theory to the more general context where all the chromosomes are general DAGs rather than total orders [37,38]. The rearrangement problem becomes: TO INFER A TRANSFORMATION SEQUENCE (TRANSLOCATIONS AND/OR REVERSALS) FOR TRANSFORMING A SET OF LINEARIZATIONS (TOPOLOGICAL SORTS), ONE FOR EACH CHROMOSOMAL DAG IN THE GENOME OF ONE SPECIES, TO A SET OF LINEARIZATIONS OF THE CHROMOSOMAL DAGS IN THE GENOME OF ANOTHER SPECIES, MINIMIZING THE NUMBER OF TRANSLOCATIONS AND REVERSALS REQUIRED.

A DAG can generally be linearized in many different ways, all derivable from a topological sorting routine. All the possible adjacencies in these linear sorts can be represented by the edges of a directed graph (DG) containing all the edges of the DAG plus two edges of opposite directions between all pairs of vertices, which are not ordered by the DAG. This is illustrated on the right in Figure 1.

We can make a bicoloured graph from the set of edges in the DGs for two partially ordered genomes. In the resulting graph, each of the DAG edges and both of the edges connecting each of the unordered pairs in the DG for each chromosome represent potential adjacencies in our eventual linearization of a genome. The n genes or markers and $2k$ dummies determine $2n + 2k$ vertices and the potential adjacencies determine the red and black edges, based on the polarity of the genes or markers. Where the construction for the totally ordered genomes contains exactly $n + k$ red edges and $n - h$ black edges, in our construction in the presence of uncertainty there are more potential edges of each colour, but only $2n + k - h$ can be chosen in our construction of the cycle graph, which is equivalent to the simultaneous linearization by topological sorting of each chromosome in each genome. IT IS THIS PROBLEM OF SELECTING THE RIGHT SUBSET OF EDGES THAT MAKES THE PROBLEM DIFFICULT (AND, WE CONJECTURE, NP-HARD.)

Our approach to this problem is a depth-first branch and bound search in the environment of $h + k$ continually updated partial orders, one for each chromosome in each genome. The strategy is to build cycles and paths one at a time. After each one is completed, the current best construction serves as a bound to compare against the maximum number of cycles and paths that could possibly be built with the remaining eligible edges. The effect of the current bound becomes greater every time a potential edge is chosen for the graph, because this generally makes many other edges ineligible to be chosen at later steps. This is not just a question of avoiding multiple edges of the same colour incident to a single vertex, but also combinations of edges that are incompatible with one of the DAGs.

We have focused here on obtaining the cycle decomposition; this is equivalent to optimally linearizing the partial orders, so that finding the rearrangements themselves can be done using the previously available algorithms and software, e.g., GRIMM [33].

One problem we have not dealt with is **conflict**; different maps of the same genome do occasionally conflict, either because $b < a$ in one data set while $a < b$ in the other or because a gene is assigned to different chromosomes in the two data sets. There are a variety of possible ways of resolving order conflicts or, equivalently, of avoiding any cycles in the construction of the DAG. One way is to delete all order relations that conflict with at least one other order relation. Another is to delete a minimal set of order relations so that all conflicts can be resolved. Perhaps the approach that best balances loss of information with ease of application and interpretation is to discard a minimum set of gene occurrences so that all order conflicts are resolved. This method also resolves conflicts due to gene assignment to different chromosomes. Any gene that is discarded from all the data sets for one genome has, of course, to be discarded from the other.

6   Synthesis and Application

Given a map comparison that suffers from some combination of **coarseness**, **missing data**, **no sign** and **paralogy**, we can ask: simultaneously find the exemplars and sign assignments resulting in a minimum number of translocations and inversions necessary to transform some DAG linearisation of one genome into some DAG linearisation of the other. Since all three component problems are hard, there is scant hope that their combination is tractable. In this section, we describe a practical approach to one problem of this type.

Note that if there is **conflict**, we might want to avoid discarding exemplars in resolving conflict; if that is impossible, then we should at least take into account the sizes of any discarded gene families in assuring a minimum of genes occurrences are discarded. In any case, this minimum should be established beforehand, and should constrain the exemplar selection, if this is an issue. Under this one constraint, the goal is the minimization of genomic distance over all combinations of exemplar choices, eligible conflict resolutions, sign assignments and DAG linearisations,

The particular application we study, using the implementation of the DAG linearisation described in [38], is the comparison of the maize and sorghum genomes. We used one set of genomic markers for maize [25] and two for sorghum [18,5] as accessible in Gramene [36]. We extracted all markers registered as having homologs in maize and at least one of the sorghum datasets. This gave 463 marker occurrences in maize and 387 in sorghum, based on 296 total non-homologous markers.

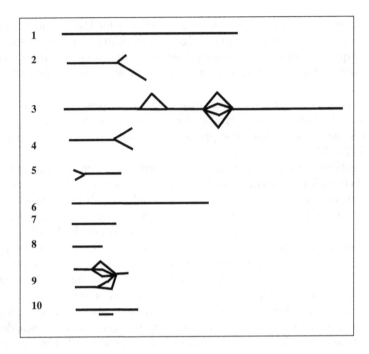

Fig. 2. DAGs for 10 sorghum chromosomes, scaled by number of markers analysed

Partly because this size of this problem is excessive for our implementation, we ignored any pair of chromosomes, one in maize and one in sorghum, with less than four markers in common. Some threshold, though perhaps not as large as four, is also justified by the facts that occasional syntenies of this sort are often the result of marker homology assignment or other error, and that especially in the case of singletons, the rearrangement solution simply includes two or three rearrangements solely to account for the position of this marker, and is independent of the rearrangement of the rest of the genome. This step left us with 381 marker occurrences in maize and 301 in sorghum, based on 263 total non-homologous markers. Thus by removing only 11% of the non-homologous markers from the original data, we remove 65 % of the excess paralogs, consistent with our suspicion that these do not represent orthologies.

As a next step, we identified all strips, as this is crucial not only to solving **no signs**, but is also helpful for **paralogy** and **conflict**. To take further advantage of strips, we removed paralogs and markers involved conflicts whenever they interrupted contiguous strips. We then found the exemplars for the remaining paralogies and resolved the remaining conflicts. To further reduce the size of the problem, we discarded a number of other singletons.

The remaining markers in the two sorghum and one maize datasets, representing 191 different markers, organized into 99 strips and singletons, could then be input into our exact linearisation algorithm. The DAGs for the sorghum chromosomes are illustrated in Figure 2. The solution involved 6 non-trivial

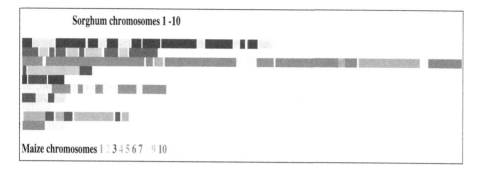

Fig. 3. Conserved segments on sorghum chromosomes, scaled by number of markers analysed

cycles (more than two edges) and 20 paths, implying a total of 73 inversions and translocations. Figure 3 portrays the configuration of the conserved segments in the two genomes, disposed on the sorghum chromosomes.

7 Discussion

A generally usable algorithm for the simultaneous solution of the linearisation and sign assignment problems seems feasible, since both can be handled within the partial order framework, though of course this is still a worst-case hard problem. There are many approaches possible to improve the current bound, to find a better sequence of edges as candidates to add to the current alternating colour cycle, and to incorporate heuristics, such as formalizing our strip-maximization/singleton-minimization procedure for discarding the most likely erroneous markers.

The situation with paralogy and conflict is more complicated, as the strong constraint of acyclicity in the DAG representation of the map data cannot be satisfied. Nevertheless, there is hope for some method drawn from the homology assignment literature we have cited to be incorporated into the solution of these problems in the comparative map context. The maize genome is known to have originated in a genome doubling event [13]; thus the treatment of duplicates through the exemplar or similar paradigm may be less appropriate than a genome halving analysis [12], which is only of polynomial complexity.

Acknowledgements

Research supported in part by grants from the Natural Sciences and Engineering Research Council of Canada (NSERC). DS holds the Canada Research Chair in Mathematical Genomics and is a Fellow of the Evolutionary Biology Program of the Canadian Institute for Advanced Research. Thanks to Glenn Tesler for a discussion of strategies in his GRIMM software.

References

1. Blin, G. and Rizzi, R. 2005. Conserved interval distance computation between non-trivial genomes. In Wang, L. (ed), Proceedings of COCOON '05. LNCS 3595. Berlin, Heidelberg:Springer Verlag, in press.
2. Bourque, G., Pevzner, P.A. and Tesler, G. 2004. Reconstructing the genomic architecture of ancestral mammals: lessons from human, mouse, and rat genomes. Genome Research, 14:507–516.
3. Bourque, G., Yacef, Y., and El-Mabrouk, N. 2005. Maximizing synteny blocks to identify ancestral homologs. manuscript.
4. Bourque, G., Zdobnov, E., Bork, P., Pevzner, P. and Tesler, G. 2005. Comparative architectures of mammalian and chicken genomes reveal highly variable rates of genomic rearrangements across different lineages. Genome Research, 15:98–110.
5. Bowers, J. E., Abbey, C., Anderson, S., Chang, C., Draye, X., Hoppe, A. H., Jessup, R., Lemke, C., Lennington, J., Li, Z., Lin, Y. R., Liu, S. C., Luo, L., Marler, B. S., Ming, R., Mitchell, S.E., Qiang, D., Reischmann, K., Schulze, S. R., Skinner, D. N., Wang, Y. W., Kresovich, S., Schertz, K. F., Paterson, A. H. 2003. A high-density genetic recombination map of sequence-tagged sites for sorghum, as a framework for comparative structural and evolutionary genomics of tropical grains and grasses. Genetics 165:367–86.
6. Bryant, D. 2000. The complexity of calculating exemplar distances. In Sankoff, D. and Nadeau, J. (eds), Comparative Genomics. Dordrecht, NL: Kluwer, pp. 207–212.
7. Caprara, A. 1997. Sorting by reversals is difficult. in Istrail, S., Pevzner, P.A. and Waterman, M.S. (eds.) Proceedings of the First Annual International Conference on Computational Molecular Biology (RECOMB'97), ACM Press, pp. 75–83.
8. Caprara, A., Lancia, G. and Ng, S.K. 2001. Sorting permutations by reversals through branch-and-price. INFORMS Journal on Computing 13, 224–244.
9. Chen, X., Zheng, J., Fu, Z., Nan, P., Zhong, Y., Lonardi, S. and Jiang, T. 2005. Assignment of orthologous genes via genome rearrangement. IEEE/ACM Transactions on Computational Biology and Bioinformatics (TCBB), in press.
10. Christie, D. A. 1996. Sorting permutations by block interchanges. Information Processing Letters 60:165–169.
11. Friedberg, R., Attie, O. and Yancopoulos, S. 2005. Efficient sorting of genomic permutations by translocation, inversion and block interchange. Bioinformatics, in press
12. l-Mabrouk, N. and Sankoff, D. 2003. The reconstruction of doubled genomes. SIAM Journal on Computing 32:754–792.
13. Gaut, B. S. and Doebley, J.F. 1997. DNA sequence evidence for the segmental allotetraploid origin of maize. Proc. Natl. Acad. Sci. U S A. 94:6809–6814.
14. Hannenhalli, S. and Pevzner, P.A. 1995. Transforming cabbage into turnip (polynomial algorithm for sorting signed permutations by reversals). In Proc. 27th Annual ACM Symposium on the Theory of Computing, pp. 178–189.
15. Hannenhalli, S. and Pevzner, P.A. 1995. Transforming men into mice (polynomial algorithm for genomic distance problem. Proceedings of the IEEE 36th Annual Symposium on Foundations of Computer Science. 581–92.
16. Hannenhalli, S. and Pevzner, P.A. 1996. To cut or not to cut (applications of comparative physical maps in molecular evolution). In Proceedings of the 7th annual ACM-SIAM symposium on discrete algorithms, Philadelphia: SIAM, pp. 304–313.
17. Kececioglu, J. and Sankoff, D. 1993. Exact and approximation algorithms for the inversion distance between two permutations. In Proceedings of 4th Combinatorial Pattern Matching symposium, LNCS 684, Springer Verlag, pp. 87–105. (Cf. Algorithmica 13: 180–210, 1995.).

18. Menz, M. A., Klein, R. R., Mullet, J. E., Obert, J. A., Unruh, N.C., and Klein, P. E. 2002. A high-density genetic map of Sorghum bicolor (L.) Moench based on 2926 AFLP, RFLP and SSR markers. Plant Molecular Biology 48:483–99.
19. Morgan, T. H., Sturtevant, A. H. , Muller, H . J., and Bridges, C.B. 1915. The mechanism of Mendelian heredity. New York: Henry Holt. and Co.
20. NCBI Human Mouse Homology.http://www.ncbi.nlm.nih.gov/Homology/
21. Nicholas, F.W., Barendse, W., Collins, A., Darymple, B.P., Edwards, J.H., Gregory, S., Hobbs, M., Khatkar, M.S., Liao, W., Maddox, J.F., Raadsma, H.W. and Zenger K. R. 2004. Integrated maps and Oxford grids: maximising the power of comparative mapping. Poster at International Society of Animal Genetics.
22. Nguyen, C.T., Tay, Y.C. and Zhang, L. 2005. Divide-and-conquer approach for the exemplar breakpoint distance. Bioinformatics, in press.
23. Palmer, J.D., Osorio, B. and Thompson, W.F. 1988. Evolutionary significance of inversions in legume chloroplast DNAs. Current Genetics 14:6574.
24. Pevzner, P.A. and Tesler, G. 2003. Human and mouse genomic sequences reveal extensive breakpoint reuse in mammalian evolution. Proc Natl Acad Sci USA 100:7672–7
25. Polacco, M.L.; Coe, E., Jr. 2002. IBM Neighbors: A Consensus Genetic Map. (http://www.maizegdb.org/ancillary/IBMneighbors.html)
26. Radcliffe, A.J., Scott, A.D. and Wilmer, R.E. 2005. Reversals and transpositions over finite alphabets. SIAM Journal on Discrete Math, in press.
27. Sankoff, D. 1999. Genome rearrangement with gene families. Bioinformatics, 15:909–917.
28. Sankoff, D. and El-Mabrouk, N. 2000. Duplication, rearrangement and reconciliation. In Sankoff, D. and Nadeau, J. H., (eds) Comparative Genomics. Dordrecht, NL: Kluwer.
29. Sankoff, D., Leduc, G., Antoine, N., Paquin, B. Lang, B.F. and Cedergren, R. 1992. Gene order comparisons for phylogenetic inference: Evolution of the mitochondrial genome. Proc. Natl. Acad. Sci. USA, 89:6575–6579.
30. Sturtevant, A. H. 1913 The linear arrangement of six sex-linked factors in Drosophila, as shown by their mode of association. Jour. Exp. Zool. 14: 43–59.
31. Sturtevant, A. H. 1921. Genetic studies on Drosophila simulans. II. Sex-linked group of genes. Genetics 6: 43–64.
32. Tang, J. and Moret, B.M.E. 2003. Phylogenetic reconstruction from gene rearrangement data with unequal gene contents. In WADS '03. LNCS 2748, Springer Verlag, pp. 37–46.
33. Tesler, G. 2002 GRIMM: genome rearrangements web server. Bioinformatics, 18, 492–3.
34. Tesler, G. 2002. Efficient algorithms for multichromosomal genome rearrangements. Journal of Computer and System Sciences 65:587–609.
35. UCSC Genome Browser
36. Ware, D., Jaiswal, P., Ni, J., Pan, X., Chang, K., Clark, K., Teytelman, L., Schmidt, S., Zhao, W., Cartinhour, S., McCouch, S. and Stein, L. 2002. Gramene: a resource for comparative grass genomics. Nucleic Acids Research 30, 103–5.
37. Zheng, C., Lenert, A. and Sankoff, D. 2005. Reversal distance for partially ordered genomes. Bioinformatics 21, in press
38. Zheng, C. and Sankoff, D. 2005. Genome rearrangements with partially ordered Chromosomes. In Wang, L. (ed), Proceedings of COCOON '05. LNCS 3595. Berlin, Heidelberg:Springer Verlag, in press.

Very Low Power to Detect Asymmetric Divergence of Duplicated Genes

Cathal Seoighe and Konrad Scheffler

University of Cape Town, Rondebosch, Cape Town 7700, South Africa
cathal@science.uct.ac.za

Abstract. Asymmetric functional divergence of paralogues is a key aspect of the traditional model of evolution following duplication. If one gene continues to perform the ancestral function while the other copy evolves a new function then we might expect a period of accelerated sequence evolution following duplication in one of the copies. In keeping with this prediction, many individual examples of asymmetric divergence at the level of protein function have been observed that are accompanied by asymmetric divergence at the sequence level. While several large-scale studies suggest that asymmetric divergence is common across a range of different organisms the degree to which they can be considered to provide an accurate estimate of its prevalence and therefore of the importance of this mode of divergence depends on both the accuracy and power of the methods that have been used. We investigated two methods that can be used to detect asymmetric duplicates using simulated data and real data from *Arabidopsis thaliana*. One of the methods detects departure from a local molecular clock for amino acid sequences and has been used previously. The second method is novel and tests for different selective constraints along the duplicated lineages using codon models of evolution. This approach is less prone to false positive results but has lower power than the molecular clock method. We find that the power to detect asymmetric divergence is low with both methods unless the effect is strong and report a surprising lack of strong evidence for asymmetric divergence in paralogues derived from the most recent round of genome duplication in *Arabidopsis*.

1 Introduction

Whether or not duplicated genes evolve at the same rate following gene duplication has important implications for models of how gene duplication contributes to the evolution of functional novelty. Arguably, asymmetric evolution following duplication lends some support to neofunctionalization models that assert that one gene of a duplicate pair often continues to perform the ancestral function, while the duplicate becomes free to evolve novel functionality [1]. Several models have been proposed that do not necessarily imply asymmetry. Both members of a duplicated pair could evolve under relaxed constraint because of redundancy and reduced purifying selective pressure [2]. The subfunctionalisation model [3,4], proposes that the protein function [3] and/or activity [4] of the ancestral gene are partitioned among the daughter genes. Most recently, hybrid models have been

A. McLysaght et al. (Eds.): RECOMB 2005 Ws on Comparative Genomics, LNBI 3678, pp. 142–152, 2005.

proposed by He and Zhang [5] and Rastogi and Liberles [6] in which subfunctionalization, which may be symmetric or asymmetric, occurs rapidly followed by neofunctionalization on a longer timescale.

Although asymmetry is not inconsistent with subfunctionalisation models [5] several examples of asymmetric divergence that appear to be associated with evolution of novel function through some kind of neofunctionalization process have been reported [7,8] and several (though not all e.g. [2,9]) previous studies have found high proportions of asymmetric paralogues in a wide range of organisms [8,10,11]. Individual examples of accelerated paralogue evolution can provide valuable insights into the function and the evolution of specific proteins and an estimate of the extent to which duplicated genes diverge asymmetrically can inform our understanding of how genomes evolve.

Methods that have been developed to detect asymmetric divergence of duplicates typically differ in the type of molecular sequences used (nucleotides, amino acids or codons), in the approach to identifying homologous sequences and in the criteria for identifying asymmetry. Several methods use triplets of sequences that include a duplicated pair and an outgroup sequence, either from the same organism [10] or from a related organism [8,9,11]. Given a pair of duplicated genes and an outgroup, duplicate genes that depart from clock-like evolution can be detected using a relative rate test (e.g. [2,9,12]). These methods require a rooted phylogenetic tree and are highly prone to false positive inference when an incorrect outgroup is used (Fig. 1b,c). Using protein sequences of duplicated yeast genes, Kellis et al. [8] uncovered several interesting examples of asymmetric divergence among the remnants of genome duplication in *Saccharomyces cerevisiae*. In this case and in several earlier studies duplicate pairs were identified on the basis of both sequence similarity as well as genomic position and the outgroup was derived from a species that diverged from *S. cerevisiae* prior to genome duplication. However, even in studies such as this that make use of positional information to identify paralogues some of the examples of asymmetry detected may still be due to incorrect phylogeny. Incorrect choice of outgroup could result, for example, from ancient tandem duplication predating both speciation and the genome duplication event and subsequent loss of different members of the tandem pair in the two genomic regions created during genome duplication. In a study in which a relatively small proportion of asymmetric paralogues is detected even a small percentage of incorrect outgroup sequences could account for a substantial proportion of the detected examples because each incorrect outgroup can have a high probability of yielding a misleading result (Fig. 1c).

Conant and Wagner [10] used a codon-based method to infer asymmetry by estimating models that included either tied or separated parameters for the rate of nonsynonymous substitution (K_a) along the branches of a phylogenetic tree leading to the duplicated genes. Their method is similar to methods that measure departure from clock-like evolution of amino acid sequences although it does have some advantages over amino acid methods. For example, it distinguishes naturally between amino acid replacements that require different numbers or types of nucleotide substitutions. However, unlike some methods that are based

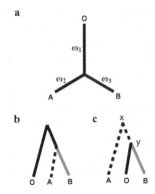

Fig. 1. (a) The tree topology used to generate the simulated data sets. A and B are paralogues and O is an outgroup. The branch leading to the outgroup was 50% longer than the branch leading to the unaccelerated paralogue. (b) Relative rate tests for asymmetry using triplets of sequences compare sequence divergence against the assumption of a local molecular clock using a rooted tree. (c) Incorrect outgroup identification will result in comparison of the dashed branch to the grey branch and will normally result in rejection of the null hypothesis (symmetric divergence) provided that the time between point x and y is sufficiently long.

on clock-like evolution of amino acid sequences it does not take account of amino acids properties. Because it measures nonsynonymous divergence this method also requires a rooted phylogeny.

In this paper we discuss an alternative codon-based method, which, as far as we are aware, has not previously been used to detect asymmetric paralogue divergence. In our approach the likelihood of a model in which ω, the ratio of nonsynonymous to synonymous substitution rates, is allowed to vary is compared to the likelihood of a model in which ω is constrained to be equal along the branches leading to the duplicated genes. Because it is based on an estimate of selective constraint, which is relatively independent of divergence, this method can accommodate different rates of mutation in different genomic regions, provides an estimate of the selective constraint along specific lineages and can be less sensitive to incorrect outgroup assignment. This is so because it does not require a rooted tree. Since there is only one unrooted tree topology for three sequences, incorrect tree topology is no longer possible. Nonetheless, this method is still vulnerable to some extent to incorrect outgroup assignment for two reasons. First, the hypothesis being tested depends on correct identification of the outgroup. If the outgroup is incorrect then a positive result could be obtained even if both copies of the gene experienced a period of relaxed selection following duplication. Secondly, ω tends to be underestimated for longer branches if there is inadequate correction for site to site rate variation [13] and this could cause the longer branch leading to one of the duplicates (Fig. 1c) to appear to have a lower rate of evolution. In this study we investigated the power of this codon-based method and compared it with the power of molecular clock methods.

2   Results

2.1   Inferring Asymmetry Using Codon Models

We simulated data sets of sequences according to the tree in Fig. 1a. Our simulation assumes that an outgroup sequence can be unambiguously identified that diverged from the duplicate pair not long prior to duplication (assuming a molecular clock, the time between the duplication and root of the tree is one quarter of the time between the duplication and the tips). The length (number of codons) of the sequence alignments and the overall length of the tree were matched with the mean value obtained from a set of *Arabidopsis* duplicated gene pairs (see Methods). Codon models of substitution [14] implemented in the PAML package [15] allow for variation in w among lineages. For every simulated alignment we compared the likelihood of a model in which the lineages leading to the duplicated genes were constrained to have the same value of w ($w_2 = w_3$ in Fig. 1a) to the likelihood of a model which was not subject to this constraint. We found that the power to detect modest differences in the selective constraint acting on one copy of a duplicated gene pair was very low. For example, if every duplicated pair of genes in a dataset was evolving with 1.5-fold asymmetry (see Methods) then we would expect to detect this asymmetry in 15% of cases (Fig. 2). Because

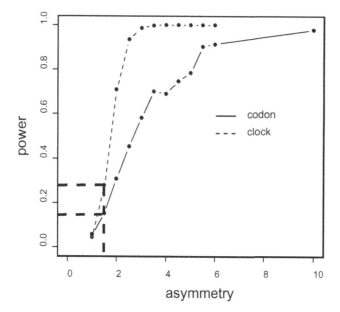

Fig. 2. Power to detect asymmetric divergence using the codon and clock methods. Each point on the graph represents the proportion of 1000 simulated data sets in which asymmetry was detected for the degree of asymmetry indicated on the x-axis. Dashed lines indicate the power to detect 1.5-fold asymmetry using the codon and clock methods.

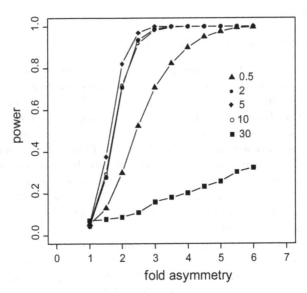

Fig. 3. Power to detect asymmetric divergence using the clock models for a range of different tree lengths. Asymmetric divergence was simulated using trees of length 30, 10, 5, 2 and 0.5 substitutions per codon.

the method we applied makes use of a 5% significance level we would need to assume that a significant proportion of the detected cases could have been the result of chance. Power to detect asymmetric divergence increases with the degree of asymmetry but does not begin to approach 100% in our simulations until we reach six-fold asymmetry (i.e. the ratio of nonsynonymous to synonymous substitution rates is six times greater in one copy than in the other).

The total tree length in the above simulations was two substitutions per codon. We simulated data sets with varying tree length to test whether our failure to detect modest asymmetric divergence was the result of a tree length which was too large (e.g. synonymous saturation on the branch leading to the outgroup sequence) or too small (insufficient time since the divergence of the duplicate pair for the asymmetry to be detected). Power to detect the asymmetry was very robust to tree length for a wide range of tree lengths (data not shown).

2.2   Inferring Asymmetry Using a Local Molecular Clock

Using the same simulated data sets described above we estimated the power of molecular-clock methods to detect asymmetric divergence in triplets of sequences. In this case we compared the likelihood of a model of evolution in which duplicated pairs of amino acid sequences evolve according to a local molecular clock to the likelihood of a model in which there was no molecular clock (see, for example Blanc and Wolfe [11]). Although the power of this method to detect modest asymmetry is still low, we found that it has far greater power than methods that are based on the ratio of nonsynonymous to synonymous substitution (Fig. 2). Again the method is quite robust to variation in the tree length (Fig. 3). Our simulations suggest that the tree lengths of the Arabidopsis duplicates and cotton outgroup (mean of 2 substitutions per codon) are close to the ideal length for detecting asymmetric divergence (Fig. 3). From approximately 3-fold asymmetry this method approaches 100% power with tree lengths in the same range as the real data, even though the codon based methods continue to miss some cases of asymmetric divergence. Although potentially more powerful than methods based on ω this method is also subject to false positives due to differences in mutation rate. Mutation rates are thought to vary substantially between genomic regions [9,16]. If one copy of the gene is located in a region of the genome with a higher mutation rate then it may have diverged much more than the other even if they are evolving under similar selective constraints. However, in this case, the increased rate of evolution in one copy does not necessarily indicate asymmetric functional divergence. Differing mutation rates between duplicates can even, under certain circumstances, contribute to long-term maintenance of genetic redundancy [17]. With both the clock-based method and the method based on ω we obtained approximately 5% false positive results, using a significance threshold of 5%. However, in real datasets false positive inference will also be affected by incorrect outgroup assignment and different mutation rates. We did not attempt to model these effects.

2.3 Asymmetric Divergence of Duplicated Genes from *Arabidopsis*

We obtained triplets consisting of *Arabidopsis* genes duplicated in the most recent round of genome duplication [18–20] and outgroup sequences from cotton (*Gossypium hirsutum*), which diverged from *Arabidopsis* prior to this genome duplication event [18,20]. We used this dataset (rather than the larger dataset of Blanc and Wolfe [11]) because it allowed us to have a similar outgroup for every duplicated pair. The distance from the outgroup may affect the power to detect asymmetric divergence [10]. From a total of 231 alignments we observed just 17 with asymmetric divergence of the duplicated genes, using the codon method and a significance level of 5%. Given the cut-off we applied we would expect approximately twelve false results from 231 tests and the observed number is, in fact, not significantly greater than this ($p = 0.15$, considering both tails of the binomial distribution). The proportion of asymmetric pairs that we detected in our *Arabidopsis* data set using the codon methods is significantly less than the frequency of asymmetric divergence observed in a set of much more ancient duplicated genes from yeast based on amino acid sequence divergence (72 from 457 [8]; $p = 0.002$) and less than the minimum proportion (20%) of asymmetric duplicate pairs observed in four yeast and animal species by Conant and Wagner [10] using a method designed to detect differences in nonsynonymous divergence of duplicates ($p = 1x10^{-6}$). It is also significantly below the estimate derived by Blanc and Wolfe [21] (173 out of 833, or 20.1%, $p = 1x10^{-6}$) using a local molecular clock for amino acid sequences. Some previous studies have found an equally low proportion of asymmetric pairs (e.g. 5% asymmetric duplicates reported by

Kondrashov et al. using a relative rate test on amino acid sequences [2]). Using the local amino acid molecular clock we detected 44 putative examples of asymmetric divergence (19%) among the *Arabidopsis* pairs, which is consistent with the proportion reported by Blanc and Wolfe [21] using a similar method. However, many of these examples could result from incorrect outgroup sequences or higher mutation rate and need not necessarily be indicative of asymmetric functional divergence or relaxed constraint on one of the copies (see Discussion).

3   Discussion

Individual examples of asymmetric evolution at the protein sequence level can provide important information about functional divergence of duplicated pairs of genes. Typically we would expect that the less diverged copy performs a function similar to the ancestral gene and that the more divergent copy either has reduced or novel functionality. In this context we are interested in the rate of change at the amino acid level. However, this rate of change can be affected by different mutation rates in different genomic regions. In contrast, the ratio of nonsynonymous to synonymous substitution rates (ω) provides a useful measure of selective constraint which is far less affected by differing mutation rates. The codon-based method that we investigate here tests whether duplicated genes evolve under asymmetric selective constraints. This asymmetric selection could result from a reduction in selective constraint and a more neutral pattern of evolution, or from positive selection acting on one copy of the gene [10].

Our results using simulated data reveal very low power to detect asymmetric evolution in duplicated genes of typical length with both methods tested, but particularly for the codon method. The power is clearly a function of the magnitude of the effect that we are trying to find and of the length of the sequence, but for modest effects (e.g. two-fold asymmetry) and average sequence lengths we are likely to fail to detect most cases using the methods investigated here (Fig. 2). In our simulations we used a single value of ω across all sites. In real data ω actually varies strongly across sites and this could increase the difficulty of detecting increases in the mean value of ω in one of the duplicates leading to a further reduction in power to detect asymmetric divergence.

Both of the methods of detecting asymmetric paralogue divergence that we investigated make use of a null hypothesis of equality (equal ω or equal rates of amino acid divergence). For closely related gene pairs this null hypothesis makes sense because if functional divergence is symmetric we would expect the sequences to evolve initially under similar selective constraints. For more divergent genes it becomes less clear that it makes sense to test against this null hypothesis. In this case the default expectation is that the evolutionary constraints under which the genes evolve remain the same over very long periods of time and the null hypothesis begins to resemble an assumption of a global molecular clock for sequences that are not accelerated by duplication. This may be a reasonable assumption for some proteins that perform core functions that do not change even on the scale of hundreds of millions of years of evolution but for many proteins this default expectation may not be justified. When asymmetric divergence is detected in very ancient duplicated pairs, how can we know that the asymmetry is actually caused by duplication? Put differently, how can we know that the departure from a common evolutionary rate is greater than occurs for most pairs of genes (for example unduplicated genes in different species)? This is normally not what is tested. In theory, we could test this by testing for different values of ω in a paralogue compared to an unduplicated orthologue but unfortunately mean values of ω may differ between species, for example because of differences in effective population sizes or demographic history.

The codon method as applied here to pairs of *Arabidopsis* duplicates combines some of the advantages of previous methods. Because we have used positional information as well as sequence information to identify triplets of duplicated genes and orthologues, we are unlikely to have a high proportion of incorrect outgroup sequences in the dataset and because the method uses an unrooted tree, the consequences of incorrect choice of outgroup are not as severe as with methods that require a rooted tree. We report a surprisingly low number of asymmetric duplicates among *Arabidopsis* paralogues from the most recent round of genome duplication using this method (not significantly greater than the number we would expect to find by chance if there are no asymmetric pairs). Using the clock based method applied by Blanc and Wolfe [11] we found 44 asymmetric pairs (19%), significantly more than with the codon based method. This should not be surprising since our simulations reveal that the codon-based methods have lower power than clock-based (relative rate) methods, possibly due to high uncertainty in the estimates of the denominator of ω.

Clock-based methods provide increased power at the cost of increased risk of false positive inference, however, because a positive result on the clock method (using nucleotide or amino acid sequences) could result from incorrect choice of outgroup or from different mutation rates in different genomic regions. From the 44 asymmetric pairs that we detected using the clock method eight had values of ω that were actually higher in the less divergent paralogue than in the more divergent one. These cases of asymmetry, at least, seem unlikely to be the result of reduced selective constraints or positive selection to evolve novel functionality in the more divergent paralogue. In general, comparing ω between the branches leading to the duplicates may provide a useful check on asymmetry detected using clock-based methods.

Power to detect asymmetric divergence is clearly also a function of sequence length. Relatively small increases in ω could be detected provided sequences are sufficiently long. For example, we found that power to detect 1.5-fold asymmetry using the codon method was approximately 40% with sequences 1,000 codons in length and 100% with sequences of 10,000 codons (data not shown). In general, power to detect cases of moderate asymmetry will depend on the ability to identify appropriate outgroup sequences with very high accuracy (so that clock-based methods can be used) and will remain low for shorter sequences. Attributing detectable asymmetry to duplication, especially for very ancient duplicates, will require more progress on the null expectation of how rates of divergence change over time in the absence of duplication.

4   Methods

4.1   Simulations

The evolverNSbranches program from the PAML package [15] was used to simulate outgroup sequences and duplicated pairs. The tree and relative lengths of branches in the simulations are shown in Fig. 1a. In the simulations the rate ratio of nonsynonymous to synonymous substitutions (ω) was the only parameter to change between the affected (i.e. accelerated) and unaffected branches. We define n-fold asymmetry as an increase by a factor of n in the value of ω along the branch leading to one of the paralogous sequences. Because increasing ω along a particular branch increases the relative length of the branch (given no change in the synonymous rate) we recalculated the relative lengths of the affected and unaffected branches for each value of n that was tested. A total of 1,000 data sets were simulated for each parameter value. The sequences were 250 codons in length and all sense codons occurred with equal frequency. We did not vary ω across sites.

4.2   Inferring Asymmetry

Two methods to detect asymmetrically evolving paralogues were evaluated. Both methods were implemented using the codeml programme from the PAML package [15]. In the codon-based method, the likelihood of a model in which ω was constrained to be the same between the duplicated lineages ($\omega_2 = \omega_3$ in Fig. 1a) was compared to the likelihood of a model in which ω was estimated separately for each branch. Twice the difference in the logarithm of the likelihoods between the more general model (with three ω parameters) and the more specific model (with two ω parameters) was compared to a χ^2 distribution with one degree of freedom and the more specific model was rejected at the 5% significance level. The second method was used previously by Blanc and Wolfe to detect asymmetric *Arabidopsis* pairs [11]. This method uses amino acid (or translated codon) sequences. Similarly to the previous method likelihoods are compared between a general model in which the rates of amino acid replacement are free to vary across all branches and a special case of this model in which a local molecular clock constrains the duplicated pair to evolve at the same rate.

5   Data

Arabidopsis gene sequences and complete tentative consensus (TC) sequences [22] from 12,005 cotton genes were downloaded from TIGR (ftp://ftp.tigr.org/ pub/data/a_thaliana/ath1/SEQUENCES/ and ftp://ftp.tigr.org/pub/data/tgi/ Gossypium/, respectively). TCs were annotated as complete if they were considered to cover at least 98% of the corresponding protein sequence. A list of *Arabidopsis* genes that were duplicated in the most recent genome duplication event was downloaded from the supplementary material of Bowers et al. [20]. Triplets of *Arabidopsis* pairs and outgroup sequences from cotton were constructed by searching the cotton TCs against the complete set of *Arabidopsis* transcript sequences using BLASTN [23]. Cotton coding sequences were identified by comparison with *Arabidopsis* proteins using GeneWise [24] and inframe alignments of coding sequences were constructed using ClustalW [25] and Tranalign from the Emboss package [26]. Only triplets with at least 50 aligned amino acids were retained, resulting in 231 in-frame alignments. We tested all triplets for saturation of synonymous substitutions by comparing the likelihood of a model in which ω was constrained to be equal to a low value (0.001) to the maximum value of the likelihood when ω was unconstrained. We do not expect to reject the simpler model if the synonymous substitutions are saturated. In the real data the simpler model was rejected in every case.

References

1. Ohno, S.: Evolution by gene duplication. Springer (1970)
2. Kondrashov, F., Rogozin, I., Wolfe, K., Koonin, E.: Selection in the evolution of gene duplications. Genome Biology **3** (2002) RESEARCH0008
3. Hughes, A.: The evolution of functionally novel proteins after gene duplication. Proc Biol Sci. **256** (1994) 119–124
4. Force, A., Lynch, M., Pickett, F., Amores, A., Yan, Y., Postlethwait, J.: Preservation of duplicate genes by complementary, degenerative mutations. Genetics **151** (1999) 1531–1545
5. He, X., Zhang, J.: Rapid subfunctionalization accompanied by prolonged and substantial neofunctionalization in duplicate gene evolution. Genetics **169** (2005) 1157–1164
6. Rastogi, S., Liberles, D.: Subfunctionalization of duplicated genes as a transition state to neofunctionalization. BMC Evol Biol. **5** (2005) 28
7. Zhang, J., Rosenberg, H., Nei, M.: Positive Darwinian selection after gene duplication in primate ribonuclease genes. Proc Natl Acad Sci U S A. **95** (1998) 3708–3713
8. Kellis, M., Birren, B., Lander, E.: Proof and evolutionary analysis of ancient genome duplication in the yeast Saccharomyces cerevisiae. Nature **428** (2004) 617–624
9. Zhang, L., Vision, T., Gaut, B.: Patterns of nucleotide substitution among simultaneously duplicated gene pairs in Arabidopsis thaliana. Mol Biol Evol. **19** (2002) 1464–1473
10. Conant, G., Wagner, A.: Asymmetric sequence divergence of duplicate genes. Genome Res. **13** (2003) 2052–2058
11. Blanc, G., Wolfe, K.: Functional divergence of duplicated genes formed by polyploidy during Arabidopsis evolution. Plant Cell **16** (2004) 1679–1691
12. Hughes, M., Hughes, A.: Evolution of duplicate genes in a tetraploid animal, Xenopus laevis. Mol Biol Evol. **10** (1993) 1360–1369
13. Nembaware, V., Crum, K., Kelso, J., Seoighe, C.: Impact of the presence of paralogs on sequence divergence in a set of mouse-human orthologs. Genome Res. **12** (2002) 1370–1376
14. Goldman, N., Yang, Z.: A codon-based model of nucleotide substitution for protein-coding DNA sequences. Mol Biol Evol. **11** (1994) 725–736
15. Yang, Z.: PAML: a program package for phylogenetic analysis by maximum likelihood. Comput Appl Biosci. **13** (1997) 555–556

16. Wolfe, K., Sharp, P., Li, W.: Mutation rates differ among regions of the mammalian genome. Nature **337** (1989) 283–285

17. Nowak, M., Boerlijst, M., Cooke, J., Smith, J.: Evolution of genetic redundancy. Nature **388** (1997) 167–171

18. Blanc, G., Hokamp, K., Wolfe, K.: A recent polyploidy superimposed on older large-scale duplications in the Arabidopsis genome. Genome Res. **13** (2003) 137–144

19. Vision, T., Brown, D., Tanksley, S.: The origins of genomic duplications in Arabidopsis. Science **290** (2000) 2114–2117

20. Bowers, J., Chapman, B., Rong, J., Paterson, A.: Unravelling angiosperm genome evolution by phylogenetic analysis of chromosomal duplication events. Nature **422** (2003) 433–438

21. Blanc, G., Barakat, A., Guyot, R., Cooke, R., Delseny, M.: Extensive duplication and reshuffling in the Arabidopsis genome. Plant Cell **12** (2000) 1093–1101

22. Quackenbush, J., Cho, J., Lee, D., Liang, F., Holt, I., Karamycheva, S., Parvizi, B., Pertea, G., Sultana, R., White, J.: The TIGR Gene Indices: analysis of gene transcript sequences in highly sampled eukaryotic species. Nucleic Acids Res **29** (2001) 159–164

23. Altschul, S., Madden, T., Schaffer, A., Zhang, J., Zhang, Z., Miller, W., Lipman, D.: Gapped BLAST and PSI-BLAST: a new generation of protein database search programs. Nucleic Acids Res **25** (1997) 3389–3402

24. Birney, E., Clamp, M., Durbin, R.: GeneWise and Genomewise. Genome Res **14** (2004) 988–995

25. Thompson, J., Higgins, D., Gibson, T.: CLUSTAL W: improving the sensitivity of progressive multiple sequence alignment through sequence weighting, position-specific gap penalties and weight matrix choice. Nucleic Acids Res **22** (1994) 4673–4680

26. Rice, P., Longden, I., Bleasby, A.: EMBOSS: the European Molecular Biology Open Software Suite. Trends Genet **16** (2000) 276–277

A Framework for Orthology Assignment
from Gene Rearrangement Data

Krister M. Swenson, Nicholas D. Pattengale, and B.M.E. Moret

Department of Computer Science,
University of New Mexico,
Albuquerque, NM 87131, USA
{kswenson, nickp, moret}@cs.unm.edu

Abstract. Gene rearrangements have been used successfully in phylogenetic reconstruction and comparative genomics, but usually under the assumption that
all genomes have the same gene content and that no gene is duplicated. While
these assumptions allow one to work with organellar genomes, they are too restrictive for nuclear genomes. The main challenge in handling more realistic data
is how to deal with gene families, specifically, how to identify orthologs. While
searching for orthologies is a common task in computational biology, it is usually
done using sequence data. Sankoff first addressed the problem in 1999, introduc-
ing the notion of exemplar, but his approach uses an NP-hard optimization step
to discard all but one member (the exemplar) of each gene family, losing much
valuable information in the process. We approach the problem using all available
data in the gene orders and gene families, provide an optimization framework in
which to phrase the problem, and present some preliminary theoretical results.

1 Introduction

Gene rearrangements have been used in phylogenetic reconstruction and comparative
genomics (see, e.g., [17,23]), but usually under the assumption that all genomes have
the same gene content and that no gene is duplicated. These assumptions allow one to
work with organellar genomes [2–5, 9, 10, 15, 21, 26], but are too restrictive for nuclear
genomes [11], where the main challenge is how to deal with gene families, specifically,
how to identify orthologs.

While searching for orthologies is a common task in computational biology, it is
usually done using sequence data; we approach that problem using gene rearrangement
data. Sankoff [19] first addressed this problem, proposing to identify within each gene
family an *exemplar* (a single gene, presumably the "original" one within that family)
and to discard all other homologs, thereby reducing the problem to one in which no gene
is duplicated. He further proposed that, for a pair of genomes, the exemplars should be
selected so as to minimize the distance (measured in terms of breakpoints or inversions)
between the two reduced genomes. One problem with this approach is that identifying
the exemplars is itself NP-hard, even when one genome contains no duplicate genes
[6]; another is that, by discarding all homologs, much valuable information is lost. (The
different numbers and arrangements of homologs need to be explained with a suitable
sequence of duplications, losses, and inversions, none of which appears in the exemplar

A. McLysaght et al. (Eds.): RECOMB 2005 Ws on Comparative Genomics, LNBI 3678, pp. 153–166, 2005.

framework.) Nguyen et al. [18] proposed a divide-and-conquer approach to compute an exemplar-based distance between two genomes in reasonable time.

Sankoff *et al.* [22] also proposed a simple heuristic based on breakpoints [20] that adds new genes incrementally at random; that heuristic performed well on a small collection of mitochondrial genomes with widely divergent contents. However, that method cannot handle duplications, only deletions and nonduplicating insertions; it is thus well suited to organellar genomes, but not to nuclear genomes, where large gene families are common. El-Mabrouk later gave an exact solution for that problem (but with respect to inversion distances), as well as a bounded-ratio approximation when both deletions and non-duplicating insertions are allowed [12]. She also developed an approach, based on her earlier work with doubled genomes, that uses both inversions and duplications [13]. Our group provided an alternate approach in which a correspondence is established between gene families on the basis of conserved segments [16,25] before completing the sequence using El-Mabrouk's algorithm; our results suggested that considering all members of a gene family yields better results than keeping only exemplars, but were limited in that the assignment of orthologs did not take into account any rearrangement structure beyond conserved segments. Chen *et al.* [8] gave a first attempt at using rearrangements and keeping more than just exemplars.

In this paper, we extend these approaches by providing an optimization framework derived from the breakpoint graph (the structure behind the last decade of work in gene rearrangements [14]) in which to phrase the problem; we give preliminary theoretical results in support of our framework.

2   Preliminaries

We are given a set of gene families S (the set of "names" of the gene families) and two genomes, G_1 and G_2. Each genome is represented as a (linear or circular) sequence of elements of S (an element may occur zero, one, or many times within the sequence), each with an associated sign (which denotes which strand the gene lies on). In this formulation, each genome consists of a single chromosome; however, the unichromosomal version embodies the heart of the orthology assignment problem and, as shown by Tesler [27] in the context of equal gene contents, a multichromosomal version does not introduce insurmountable problems. The problem is to find the shortest *edit sequence*, that is, the shortest sequence of evolutionary events that transforms one genome into the other. Permitted evolutionary events in this setting are *inversions*, which take a subsequence of genes and reverse it in place (in both order and signs), *deletions*, and *insertions* (including *duplications*). These events all operate on consecutive subsequences of genes: that is, we assume that the cost of deleting, inserting, or making one duplicate of, one gene is the same as that of deleting or inserting (including duplicating insertions) a contiguous segment of k genes, for any $k \geq 1$.

In absence of other constraints, the edit distance between any two genomes is then bounded by 2: simply delete the entire genome in one operation of unit cost, then insert the entire new genome in another operation of unit cost. Since this scenario is patently absurd in biological terms, we impose a simple parsimony constraint on any editing scenario: if G_1 has a family of k_1 genes and G_2 a homologous family of k_2 genes, with $k_1 \geq k_2$, then none of the k_2 genes in G_2's family may be inserted in the edit sequence from G_1 to G_2: instead, we must identify within G_1's family of k_1 genes a distinct ortholog for each of the k_2 genes in G_2's family. The $k_1 - k_2$ unmatched homologs in G_1's family will then be deleted in the edit sequence. Once that orthology identification has been made, the algorithms of El-Mabrouk [12] and of our group [11,26] can complete the work of finding one or more parsimonious edit sequences.

A good choice of orthologies can reduce the required number of deletions and insertions (or duplications) by inserting contiguous segments of many genes rather that one gene at a time—although care must be taken not to do so at the expense of proper placement within the ordering, lest many additional inversions be required to move individual genes to their final destination [16]. It can also reduce the number of required inversions by grouping genes properly: this is the focus of the cover-based methods [8,16,25].

We shall rely on the fact that every gene that appears as a singleton in both genomes has a direct assignment and that these singleton genes must all be sorted through inversions: because we know how to sort by inversions [1,14], the presence of singleton genes creates a structural context in which to study orthology assignment [19].

We assume, without loss of generality, that gene families present in one genome but not the other have been removed—these families do not affect orthology assignment and the insertion of the unique genes can easily be handled by El Mabrouk's algorithm. We describe the framework for the general case, but, for the sake of clarity in presentation, we shall frequently restrict genome G_2 to contain no duplicate genes, in which case our framework becomes a special case of the exemplar problem. Finally, when using G_2 with no duplicate genes, we assume that the remaining genes have been indexed from 1 to n so as to turn G_2 into the identity permutation $12 \ldots n$. (As is customary, we will prepend a marker 0^+ and append another marker $n + 1^-$ to both genomes.)

3 Background and Definitions

3.1 The Breakpoint Graph

The basic structure describing a pair of genomes with no duplicates and equal gene content is the *breakpoint graph* (really a multigraph)—for a readable description of its construction, see [24]. In our case, however, gene families need not be singletons, so we modify the construction to include *only* singleton gene families. Let $BG_{1,2}$ denote the breakpoint graph for G_1 and G_2; As in the normal breakpoint graph, each singleton gene g becomes a pair of vertices, g^- and g^+ (the "negative" and "positive" terminals); however, we leave out the gene families with multiple members, since only the singletons have a readily usable structure. We need to accommodate gaps left in the sequence where duplicate genes exist in G_1. Call the versions of G_1 and G_2 without multigene families G_1' and G_2' respectively. We add an edge (a *desire* edge, in the charming terminology of [24], but also known elsewhere as a gray edge) (a^-, b^+) for each singleton a and b, whenever a occurs immediately to the left of b in G_2'. We add a *reality* edge (also known elsewhere as a black edge) (a^p, b^q) if a is the element to the left of b in G_1' and we have either $p = q$ if a and b have different parities (in G_1', naturally) or $p \neq q$ if a and b have the same parity. Thus desire edges trace the (re-)ordering of G_1 that we need to achieve to match G_2, while reality edges trace the given ordering of G_1. Figure 1 illustrates the construction.

$$G_1 = 4 \text{ -}3\ 2\ 3\ 1\ 6\ 9\ 3\ 8 \text{ -}10 \text{ -}7\ 9$$

(a) the genome G_1

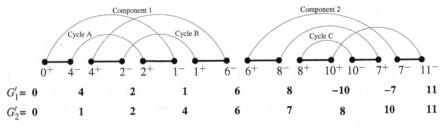

(b) the breakpoint graph $BG_{1,2}$

Fig. 1. A genome G_1 and its associated breakpoint graph $BG_{1,2}$ (with respect to the identity permutation G_2) after removing gene families with duplicates (3 and 9); desire edges are shown in gray, reality edges in black

Hannenhalli and Pevzner proved that the inversion distance equals the number of genes minus the number of cycles in the breakpoint graph, plus some corrective factors (hurdles and a fortress). Researchers have found that hurdles are very rare in real data (a finding confirmed in a theorem under some restrictive assumptions [7]), so we focus on selecting an orthology assignment that maximizes the number of cycles.

3.2 The Consequences of An Assignment

Our job of assigning orthologs may be compared to that of reshelving books in a library with unlabelled shelves. Each book has a proper location on a shelf and multiple copies of a book must be shelved together. A librarian can proceed by first removing misshelved books and then identifying the appropriate location of each book based on the context of the books that remain in their correct spot.

In our problem each multigene family has been removed from the ordering, leaving a structure of cycles defined by singleton genes. We call each gene in a multigene family of G_2 a *candidate*, since it is one of the choices for an orthology assignment to a corresponding gene in G_2. Like each book in the library, each candidate has a location between two remaining elements in G_2'; each family, like each group of book copies, contains candidates that all share the same location between elements of G_2'. For each candidate d, denote by $\beta^+(d)$ the positive terminal of the next smaller (in value) element in $BG_{1,2}$ and by $\beta^-(d)$ the negative terminal of the next larger element. We call these nodes the *bookends* of d and the cycle on which they reside the *shelf* of d. For instance, in Figure 1, the bookends for the family of gene 3 (a family of 3 members) are 2^+ and 4^- and therefore the shelf for the family of 3s is cycle A. Although the definition of bookends applies equally well to singletons, we are only interested in bookends for candidates: bookends are part of the breakpoint graph, but candidates are not, since multigene families do not appear in the breakpoint graph.

Once we have chosen a candidate, the candidate and its matching gene in G_2 effectively form a singleton gene family, so we can add the candidate to the breakpoint graph of G_1. The consequences of that choice are summarized in the following easy lemma, which underlies many of our results.

Lemma 1. *When a candidate d is chosen, exactly two edges are affected: the reality edge that spans the location where d is added and the edge between its bookends.*

Proof. Refer to Figure 2. Adding d to $BG_{1,2}$ splits the reality edge that spans the location where d is added, creating two new endpoints d^+ and d^-, as well as splitting the desire edge that links $\beta^+(d)$ and $\beta^-(d)$ to meet each of d^+ and d^-.

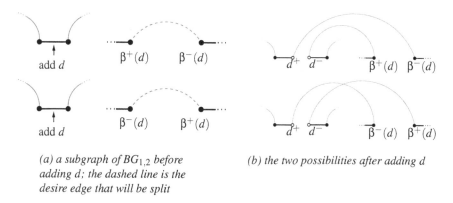

(a) a subgraph of $BG_{1,2}$ before adding d; the dashed line is the desire edge that will be split

(b) the two possibilities after adding d

Fig. 2. Adding an element d to a breakpoint graph

We say that a candidate d is added *on-cycle* if, once added, it lies on its own shelf; otherwise it is added *off-cycle*. The following is an immediate consequence of Lemma 1.

Lemma 2. *When a candidate is added off-cycle, two cycles get joined.*

3.3   The Cycle Splitting Problem

We can formulate orthology assignment as an optimization problem: choose an assignment of orthologs that maximizes the number of cycles in the resulting breakpoint graph (i.e. $BG_{1,2}$, to which the chosen candidates have been added). Note that the order in which the chosen candidates are added does not affect the structure of the resulting breakpoint graph.

Consider cycle C in Figure 1. This cycle is associated with the gene segment $(6, 9, 8, -10, -7, 9, 11)$, which contains two occurrences of gene 9; thus we must choose which of these two occurrences to call the ortholog of gene 9 in G_2. Figure 3 shows the augmented breakpoint graphs resulting from each choice of candidate. The graph on the left, where we chose the candidate between 6 and 8, has one more cycle than the graph on the right, where we chose the candidate between -7 and 11, and is thus the better choice.

The choice of a candidate is advantageously viewed on a breakpoint graph inscribed in a series of circles, one for each cycle in the graph. We embed each cycle of $BG_{1,2}$ in

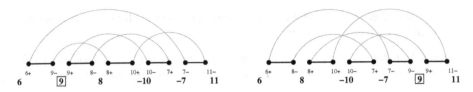

Fig. 3. The breakpoint graphs for the two candidates for gene 9 on cycle C

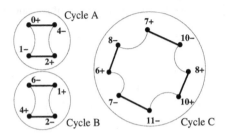

Fig. 4. The breakpoint graph of Figure 1 inscribed in three circles (cycle D is not shown)

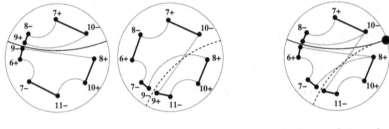

(a) the graphs of Figure 3 inscribed in circles   *(b) the two choices of part (a) superimposed*

Fig. 5. How the cycle splitting problem can be inscribed in a single circle

a circle by choosing any start vertex and then following the cycle. Figure 4 shows three of the four cycles of Figure 1 inscribed in three circles. Returning to the two possible orthology assignments shown in Figure 3, we can look at the inscribed versions of these graphs, as illustrated in Figure 5(a). Choosing candidates adds edges across the circle, edges that may cross each other, depending on the parity of the candidates and the locations of their bookends. The effects on the graph can be represented in just one graphical representation, as shown in Figure 5(b). In this representation, we denote the two choices by drawing two curved line segments, both originating on the perimeter between the bookends 10^- and 8^+ and each ending between the two terminals of the corresponding candidate. Choosing the candidate between 6^+ and 8^- gives rise to desire edges that do not cross in the inscribed representation; we represent such choices with solid lines. The other candidate, between 7^- and 11^-, does give rise to crossing desire edges; we represent such choices with dashed lines.

These curved lines represent assignment *operations*; we will call an operation represented by a solid line a *straight* operation (because it does not introduce crossings) and one represented by a dashed line a *cross* operation. The collection of all operations that share an endpoint represents all members of a gene family from G_1, so we also call it a *family* and call its common endpoint (between the bookends and represented by a large disk in the figures) the *family home*. We can now state the three constraints for our optimization problem:

1: Each family home is a distinct point on the circle.
2: The family home is not the endpoint of any operation not in that family.
3: The other endpoint of each operation is unique to that operation.

The objective to be maximized is the number of cycles. Figure 6 shows the operations for each of the gene families from our running example. Operations that cross cycles are off-cycle and therefore will join cycles.

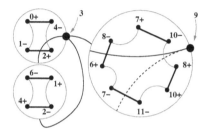

Fig. 6. The operations that represent the gene families for our running example

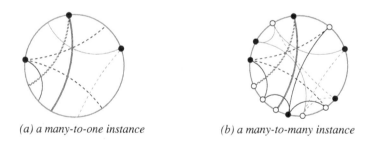

(a) a many-to-one instance   (b) a many-to-many instance

Fig. 7. A single cycle for the simplified case (left) and the general one (right)

Figure 7 shows a single cycle and its operations for the simplified ("many-to-one") case where G_2 has only singletons and for the general ("many-to-many") case where both G_1 and G_2 have multigene families. (The case where two multigene families have the same bookends can be handled because the relative location of the bookends does not change.) In the general case we have multiple homes per family, with one additional constraint:

4: Each home in the same family must connect to all of the same endpoints.

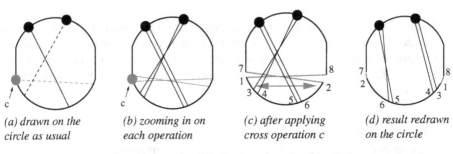

(a) drawn on the circle as usual   *(b) zooming in on each operation*   *(c) after applying cross operation c*   *(d) result redrawn on the circle*

Fig. 8. Illustration for Theorem 1. Labels for the points along the circle are numbered.

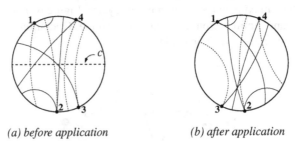

(a) before application   *(b) after application*

Fig. 9. Applying cross operation c

The problem thus becomes picking as many operations as there are homes per family such that the cycle count is maximized. The only additional complication is that applying an operation removes that operation from consideration in all other homes for its family (as required by the fourth constraint).

Straight and cross operations display a form of duality that allows us to focus on straight operations alone.

Theorem 1. *Applying a cross operation c converts all operations that intersect c (call the set of such operations I) to their complement—crosses are replaced by straights and straights by crosses. Furthermore, for any two operations in I, if they intersected before applying c, then they no longer do after applying c, and vice versa.*

Proof. We sketch the proof graphically, using Figure 8, a typical situation where three operations, two of which are crosses and one a straight, overlap each other. The cross operation shown in parts (a) and (b) twists, but does not break the cycle, as shown in part (c). If we redraw the cycle inscribed neatly in a circle, we find we must reverse the indices on half of the cycle; Figure 8(d) shows the result after reversing indices on the bottom half of the cycle. Previously intersecting operations no longer intersect and the identities of the operations have been inverted.

Figure 9 shows the implications of Theorem 1 in a more complicated setting.

4   Theoretical Results

4.1   Buried Operations

An operation makes no contribution to the cycle count of a complete assignment if the two new desire edges it creates lie on the same cycle. In Figure 10, the choices of candidates for the gene families are indicated in the breakpoint graph on the left and shown as operations in the inscribed representation on the right.

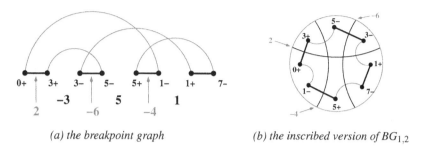

(a) the breakpoint graph   (b) the inscribed version of $BG_{1,2}$

Fig. 10. An example with $G_1 = 2, -3, 4, -6, 5, -4, -2, 6, 1$. Chosen duplicates are shown in grey.

In Figure 11, we show again the three operations depicted in Figure 10(b), but this time only the three operations and the resulting two cycles are shown. Note the operation corresponding to gene family 2 (shown as a heavy curve): the curved edge is bounded on each side by the same cycle; we say that such an operation is *buried* (for the given choice of candidates). Since the two desire edges created by this operation lie on the same cycle, the operation does not increase the number of cycles (in fact, in this particular example, it reduces the number of cycles, which stood at 3 after operations -6 and -4).

Fig. 11. The cycle and the operations; operation "2" (the heavy curve) is buried

Theorem 2. *If an orthology assignment creates a total of b buried edges, then the number of cycles is bounded by $a - b + 1$, where a is the number of cycles present in the breakpoint graph induced by the shared singleton genes plus the total number of orthology assignments to be made.*

Proof. The number of cycles cannot exceed $a + 1$, since each orthology assignment can give rise to at most one new cycle. Consider the effect on the breakpoint graph of choosing an operation: a single desire edge d is replaced with two desire edges d_1' and d_2', and a single reality edge r is replaced with two reality edges r_1' and r_2'. By construction, d_1' and d_2' each inherit one of the original endpoints of d; similarly, r_1' and r_2' each inherit one of the original endpoints of r. By assumption, the chosen edge is buried, so that d'_1 and d'_2 lie on the same cycle; therefore so do all of the original endpoints of d and r. Thus all of the newly created edges must lie on a cycle that already existed. Since this is true of any buried operation, every one of the buried operations decreases by one the maximum number of attainable cycles.

4.2 Chains and Stars

We have discovered two operation patterns that, while they need not contain buried operations, nevertheless impose sharp bounds on the number of cycles. A *k-chain* (for $k \geq 3$) is an assignment in which k operations form a chain, that is, each chosen operation overlaps two of the other k, its predecessor and successor around the circle. Figure 12(a,b) illustrates k-chains. A *k-star* (for $k \geq 1$) is an assignment in which k operations form a clique (each overlaps every other). Figure 12(c,d) illustrates k-stars.

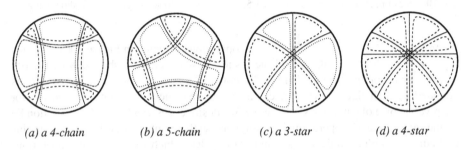

(a) a 4-chain   (b) a 5-chain   (c) a 3-star   (d) a 4-star

Fig. 12. Some examples of stars and chains

Proposition 1. *For any integer $k \geq 1$ (but recall that k-chains are only defined for $k \geq 3$), we have:*

1. *a k-chain has no buried operations;*
2. *in a k-chain with k odd, the cycle count is 2;*
3. *in a k-chain with k even, the cycle count is 3;*
4. *in a k-star with k even, every operation is buried and the cycle count is 1;*
5. *in a k-star with k odd, no operation is buried and the cycle count is 2.*

We conjecture that these two patterns, along with buried operations, describe all operations that reduce the upper bounds on the number of cycles.

4.3 Reduced Forms

A serial assignment procedure could reach a state in which no operation remains that could split a cycle. We call such a state a *reduced form* of the instance. In a reduced form, an instance is composed of multiple cycles linked by the operations from the remaining families. This structure lends itself naturally to a graph representation; an analysis of this graph reveals conditions under which optimality can be characterized.

Theorem 3. *After applying a maximal nonoverlapping set of operations M, remaining operations can only (by themselves) join two cycles.*

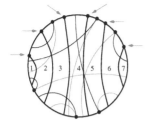

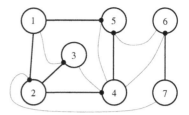

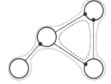

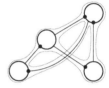

(a) operations indicated by heavy lines (and arrows) are those chosen to produce the reduced form of part (b)

(b) the resulting reduced instance; heavy edges will produce an optimal solution to the reduced instance

Fig. 13. Creating a reduced instance and solving it

(a) the effect of applying an operation between two circles

(b) a reduced form: lines trace the cycles created by the operations

(c) adding a nonplanar operation to the reduced form from (b) joins the cycles

Fig. 14. The effect of choosing operations on a reduced form

Proof. Applying a set of k nonoverlapping operations yields k new cycles, each separated from the others by two adjacent operations or, in the case of an outermost cycle, by one operation from all others. Since M is maximal, every remaining operation from every family overlaps an element of M. Application of any $m \in M$, therefore, must span two of the new cycles, joining them into one.

Figure 13(b) shows the reduced instance induced by applying each of the (straight) operations chosen in Figure 13(a). We are left with a reduced form that can be viewed as a graph on the cycles so far; however, because that graph is embedded in the plane, the edges incident on a vertex are strictly ordered.

We can now take advantage of graph properties such as planarity, cycles, and connected components. Because of the ordered nature of the edges incident upon a given vertex, planarity is somewhat specialized in our case: nonplanar edges can occur in simpler situations than in general graphs, as shown in Figure 14(c). Cycles again play a vital role in these new graphs. If we restrict our attention to planar graphs, we can look at the elementary cycles (those that delimit an inside face of the planar embedding) and obtain directly the value of an optimal solution. As shown in Figure 14, each connected component produces a cycle around its outer hull (one of the cycles for the outer face of the planar graph). Each elementary cycle yields another cycle to its inside. Figure 14(c) shows how nonplanar edges can join these two cycles.

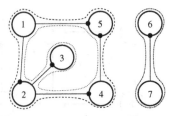
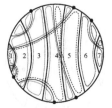

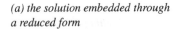

(a) the solution embedded through   (b) the solution embedded on the circle
a reduced form

Fig. 15. An optimal solution to the reduced instance in Figure 13

Theorem 4. *The number of cycles in a solution S to a planar reduced instance with m elementary cycles and c connected components is* $R(S) = m + c$.

Proof. This certainly holds for a reduced instance with no operations. Assume $R(S) = m + c$ for a current solution and look at the effect of adding another edge. If that edge links two previously disconnected components, then the cycles around the hulls of these components will get merged, removing a cycle and a connected component. If that edge links two connected components, then an elementary cycle will be created. Since the edge added is planar, we know that the same cycle runs past both endpoints of the operations and thus the operation will split it.

It remains to relate results on reduced forms back to the original inscribed breakpoint graph formulation; we illustrate the process in Figure 15, where the left part shows the solution obtained on a reduced form and the right part shows the corresponding solution inscribed in the circle.

5   Conclusion

We have described a graph-theoretical framework in which to represent and reason about orthology assignments and their effect on the number of cycles present in the resulting breakpoint graph. We have given some foundational results about this framework, including several that point us directly to to algorithmic strategies for optimizing this assignment. We believe that this framework will lead to a characterization of the orthology assignment problem as well as to the development of practical algorithmic solutions.

Note that research in orthology assignment based on rearrangement data does not aim to replace assignment based on sequence data: instead, the two approaches complement each other. Biologists are already routinely using the notion of contiguous gene blocks in their determination of orthology assignments: an assignment based on rearrangement data simply formalizes that insight. How to use both sequence data and gene-rearrangement data within the same framework remains a tantalizing, but for now elusive goal.

Acknowledgments

This work is supported by the US National Science Foundation under grants EF 03-31654, IIS 01-13095, IIS 01-21377, and DEB 01-20709, and by the US National Institutes of Health under grant 2R01GM056120-05A1.

References

1. D.A. Bader, B.M.E. Moret, and M. Yan. A fast linear-time algorithm for inversion distance with an experimental comparison. *J. Comput. Biol.*, 8(5):483–491, 2001.
2. M. Blanchette, T. Kunisawa, and D. Sankoff. Gene order breakpoint evidence in animal mitochondrial phylogeny. *J. Mol. Evol.*, 49:193–203, 1999.
3. J.L. Boore. Phylogenies derived from rearrangements of the mitochondrial genome. In N. Saitou, editor, *Proc. Int'l Inst. for Advanced Studies Symp. on Biodiversity*, pages 9–20, Kyoto, Japan, 1999.
4. J.L. Boore and W.M. Brown. Big trees from little genomes: Mitochondrial gene order as a phylogenetic tool. *Curr. Opinion Genet. Dev.*, 8(6):668–674, 1998.
5. J.L. Boore, T. Collins, D. Stanton, L. Daehler, and W.M. Brown. Deducing the pattern of arthropod phylogeny from mitochondrial DNA rearrangements. *Nature*, 376:163–165, 1995.
6. D. Bryant. The complexity of calculating exemplar distances. In D. Sankoff and J. Nadeau, editors, *Comparative Genomics: Empirical and Analytical Approaches to Gene Order Dynamics, Map Alignment, and the Evolution of Gene Families*, pages 207–212. Kluwer Academic Publishers, Dordrecht, NL, 2000.
7. A. Caprara. On the tightness of the alternating-cycle lower bound for sorting by reversals. *J. Combin. Optimization*, 3:149–182, 1999.
8. X. Chen, J. Zheng, Z. Fu, P. Nan, Y. Zhong, S. Lonardi, and T. Jiang. Computing the assignment of orthologous genes via genome rearrangement. In *Proc. 3rd Asia Pacific Bioinformatics Conf. (APBC'05)*, pages 363–378. Imperial College Press, London, 2005.
9. M.E. Cosner, R.K. Jansen, B.M.E. Moret, L.A. Raubeson, L. Wang, T. Warnow, and S.K. Wyman. An empirical comparison of phylogenetic methods on chloroplast gene order data in Campanulaceae. In D. Sankoff and J.H. Nadeau, editors, *Comparative Genomics*, pages 99–122. Kluwer Academic Publishers, Dordrecht, NL, 2000.
10. S.R. Downie and J.D. Palmer. Use of chloroplast DNA rearrangements in reconstructing plant phylogeny. In D.E. Soltis, P.S. Soltis, and J.J. Doyle, editors, *Molecular Systematics of Plants*, pages 14–35. Chapman and Hall, New York, 1992.
11. J. Earnest-DeYoung, E. Lerat, and B.M.E. Moret. Reversing gene erosion: reconstructing ancestral bacterial genomes from gene-content and gene-order data. In *Proc. 4th Int'l Workshop Algs. in Bioinformatics (WABI'04)*, volume 3240 of *Lecture Notes in Computer Science*, pages 1–13. Springer Verlag, Berlin, 2004.
12. N. El-Mabrouk. Genome rearrangement by reversals and insertions/deletions of contiguous segments. In *Proc. 11th Ann. Symp. Combin. Pattern Matching (CPM'00)*, volume 1848 of *Lecture Notes in Computer Science*, pages 222–234. Springer Verlag, Berlin, 2000.
13. N. El-Mabrouk. Reconstructing an ancestral genome using minimum segments duplications and reversals. *J. Comput. Syst. Sci.*, 65:442–464, 2002.
14. S. Hannenhalli and P.A. Pevzner. Transforming cabbage into turnip (polynomial algorithm for sorting signed permutations by reversals). In *Proc. 27th Ann. ACM Symp. Theory of Comput. (STOC'95)*, pages 178–189. ACM Press, New York, 1995.
15. B. Larget, D.L. Simon, and J.B. Kadane. Bayesian phylogenetic inference from animal mitochondrial genome arrangements. *J. Royal Stat. Soc. B*, 64(4):681–694, 2002.
16. M. Marron, K.M. Swenson, and B.M.E. Moret. Genomic distances under deletions and insertions. *Theor. Computer Science*, 325(3):347–360, 2004.

17. B.M.E. Moret, J. Tang, and T. Warnow. Reconstructing phylogenies from gene-content and gene-order data. In O. Gascuel, editor, *Mathematics of Evolution and Phylogeny*, pages 321–352. Oxford University Press, UK, 2005.
18. C. Thach Nguyen, Y.C. Tay, and L. Zhang. Divide-and-conquer approach for the exemplar breakpoint distance. *Bioinformatics*, 21(10):2171–2176, 2005.
19. D. Sankoff. Genome rearrangement with gene families. *Bioinformatics*, 15(11):990–917, 1999.
20. D. Sankoff and M. Blanchette. The median problem for breakpoints in comparative genomics. In *Proc. 3rd Int'l Conf. Computing and Combinatorics (COCOON'97)*, volume 1276 of *Lecture Notes in Computer Science*, pages 251–264. Springer Verlag, Berlin, 1997.
21. D. Sankoff and M. Blanchette. Multiple genome rearrangement and breakpoint phylogeny. *J. Comput. Biol.*, 5:555–570, 1998.
22. D. Sankoff, D. Bryant, M. Deneault, B.F. Lang, and G. Burger. Early Eukaryote evolution based on mitochondrial gene order breakpoints. *J. Comput. Biol.*, 7(3):521–536, 2000.
23. D. Sankoff and J. Nadeau, editors. *Comparative Genomics: Empirical and Analytical Approaches to Gene Order Dynamics, Map Alignment, and the Evolution of Gene Families*. Kluwer Academic Publishers, Dordrecht, NL, 2000.
24. J.C. Setubal and J. Meidanis. *Introduction to Computational Molecular Biology*. PWS Publishers, Boston, MA, 1997.
25. K.M. Swenson, M. Marron, J.V. Earnest-DeYoung, and B.M.E. Moret. Approximating the true evolutionary distance between two genomes. In *Proc. 7th SIAM Workshop on Algorithm Engineering & Experiments (ALENEX'05)*. SIAM Press, Philadelphia, 2005.
26. J. Tang, B.M.E. Moret, L. Cui, and C.W. dePamphilis. Phylogenetic reconstruction from arbitrary gene-order data. In *Proc. 4th IEEE Symp. on Bioinformatics and Bioengineering BIBE'04*, pages 592–599. IEEE Press, Piscataway, NJ, 2004.
27. G. Tesler. Efficient algorithms for multichromosomal genome rearrangements. *J. Comput. Syst. Sci.*, 65(3):587–609, 2002.

Author Index

Bachrach, Abraham   1
Blin, Guillaume   11
Bourque, Guillaume   21

Cannarozzi, Gina   61
Carmel, Liran   35
Chauve, Cedric   11
Chen, Kevin   1
Csűrös, Miklós   47

Dalpasso, Marcello   88
Dessimoz, Christophe   61
Durand, Dannie   73, 106

El-Mabrouk, Nadia   21

Fertin, Guillaume   11

Gil, Manuel   61
Gonnet, Gaston H.   61

Haque, Lani   121
Harrelson, Chris   1
Hoberman, Rose   73

Koonin, Eugene V.   35

Lancia, Giuseppe   88
Lenert, Aleksander   131

Marcus, Jeffrey M.   97
Margadant, Daniel   61
Mihaescu, Radu   1
Moret, Bernard M.E.   153

Pattengale, Nicholas D.   153

Raghupathy, Narayanan   106
Rao, Satish   1
Rizzi, Romeo   88
Rogozin, Igor B.   35
Roth, Alexander   61

Sankoff, David   121, 131
Scheffler, Konrad   142
Schneider, Adrian   61
Seoighe, Cathal   142
Shah, Apurva   1
Swenson, Krister M.   153

Wolf, Yuri I.   35

Yacef, Yasmine   21

Zheng, Chungfang   131

Lecture Notes in Bioinformatics

Vol. 3678: A. McLysaght, D.H. Huson (Eds.), Comparative Genomics. VIII, 167 pages. 2005.

Vol. 3615: B. Ludäscher, L. Raschid (Eds.), Data Integration in the Life Sciences. XII, 344 pages. 2005.

Vol. 3594: J.C. Setubal, S. Verjovski-Almeida (Eds.), Advances in Bioinformatics and Computational Biology. XIV, 258 pages. 2005.

Vol. 3500: S. Miyano, J. Mesirov, S. Kasif, S. Istrail, P. Pevzner, M. Waterman (Eds.), Research in Computational Molecular Biology. XVII, 632 pages. 2005.

Vol. 3388: J. Lagergren (Ed.), Comparative Genomics. VII, 133 pages. 2005.

Vol. 3380: C. Priami (Ed.), Transactions on Computational Systems Biology I. IX, 111 pages. 2005.

Vol. 3370: A. Konagaya, K. Satou (Eds.), Grid Computing in Life Science. X, 188 pages. 2005.

Vol. 3318: E. Eskin, C. Workman (Eds.), Regulatory Genomics. VII, 115 pages. 2005.

Vol. 3240: I. Jonassen, J. Kim (Eds.), Algorithms in Bioinformatics. IX, 476 pages. 2004.

Vol. 3082: V. Danos, V. Schachter (Eds.), Computational Methods in Systems Biology. IX, 280 pages. 2005.

Vol. 2994: E. Rahm (Ed.), Data Integration in the Life Sciences. X, 221 pages. 2004.

Vol. 2983: S. Istrail, M.S. Waterman, A. Clark (Eds.), Computational Methods for SNPs and Haplotype Inference. IX, 153 pages. 2004.

Vol. 2812: G. Benson, R.D. M. Page (Eds.), Algorithms in Bioinformatics. X, 528 pages. 2003.

Vol. 2666: C. Guerra, S. Istrail (Eds.), Mathematical Methods for Protein Structure Analysis and Design. XI, 157 pages. 2003.